Scaling Renewables: Challenges & Solutions

Naina

This chapter has outlined the key focus and objectives of the research topic. An introduction to large scale grid connected renewable energy (referred to as 'RE' in this proposal) and the framework for Australia's National Electricity Market under which RE operates in Australia is provided. An outline of the general barriers to large scale grid connected renewable energy integration has been provided as well as the supportive frameworks instituted to address such barriers, with specific reference to the Australian market.

The problem statement, research question, research proposition, aim and objectives of the study are highlighted. The methodology to be implemented for achieving the aim and objectives of the study is outlined, as well as the limitations to the study. The structure that the report shall follow is outlined and finally a concluding statement for the chapter is given.

1.2.1   Grid Integrated Renewable Energy

Renewable energy generation can refer to multiple technologies and how they are implemented. These technologies can refer to energy generated from wind, solar photovoltaic, hydropower, tidal/wave, geothermal, concentrated solar power and biomass sources (Trieb, 2010: 7; Martin and Rice, 2012). The implementation of the technology may be stand-alone supplying a user directly, mini-grid for rural electrification, or small-scale and large-scale grid connected systems. For the purpose of the research topic, renewable energy (RE) refers to large scale (as defined by the Clean Energy Regulator of Australia (CER, 2017)) grid connected wind and solar PV systems.

The proportion of RE is small in comparison to conventional sources of power. Global installed wind and solar-powered electricity capacity stood at 514 Gigawatt (GW) and 391 Gigawatt Electric (GWe) at the end of 2017 respectively (IRENA, 2018). Wind accounted for about seven percent of total power generation capacity and solar-powered electricity produced one percent of all electricity consumed, according to the World Energy Council's World Energy Resources report for 2016 (WEC, 2016). The RE industry is therefore clearly in the minority when compared to conventional fuel sources.

A common structure for the development of renewable energy projects is through the project being developed, owned and financed by private non-utility generators. The electricity is commonly sold through long-term power purchase agreements (PPA) which supports project financing which is the dominant renewable energy financing structure (Wiser, 1997; Tang and Zhang, 2019). Key stakeholders include government, the network service providers (NSP) who provide the distribution and transmission networks required to deliver the energy, the community, the project developer, electricity retailers/government businesses responsible for the off-take of energy, finance partners, and equipment suppliers (Martin and Rice, 2012).

1.2.2   Australian Energy Market

The Australian Energy Market Operator (AEMO) reports on the planned generation capacity . The information is updated every six months based on new data collected. Figure 1-1 shows the distribution of technologies contributing to electricity generation in the National Electricity Market (NEM) (AEMO, 2018b). Coal is by far the dominant source of generation, with 22,916 MW of existing capacity. This is supported by Australia's extensive coal reserves (Byrnes *et al.*, 2013). Utility scale solar has a minimal amount of existing generation at 323 MW, with wind more significant at 4,462 MW of existing capacity. Committed projects refer to those that have a formal commitment to construction, such as secured financing and have started procurement. Solar and wind demonstrate the largest share of committed projects, as well as an enormous majority of proposed projects that have been announced. The industry is clearly booming with a drastic change in the energy landscape expected in the future, should all proposed projects come to fruition. It is expected that there may be barriers to the planned integration of renewables on such a scale that may emerge. Australia's energy policy has taken significant steps to supporting the deployment of cleaner generation technologies, although hurdles at federal and state level have in the past frustrated the development of renewables (Byrnes *et al.*, 2013).

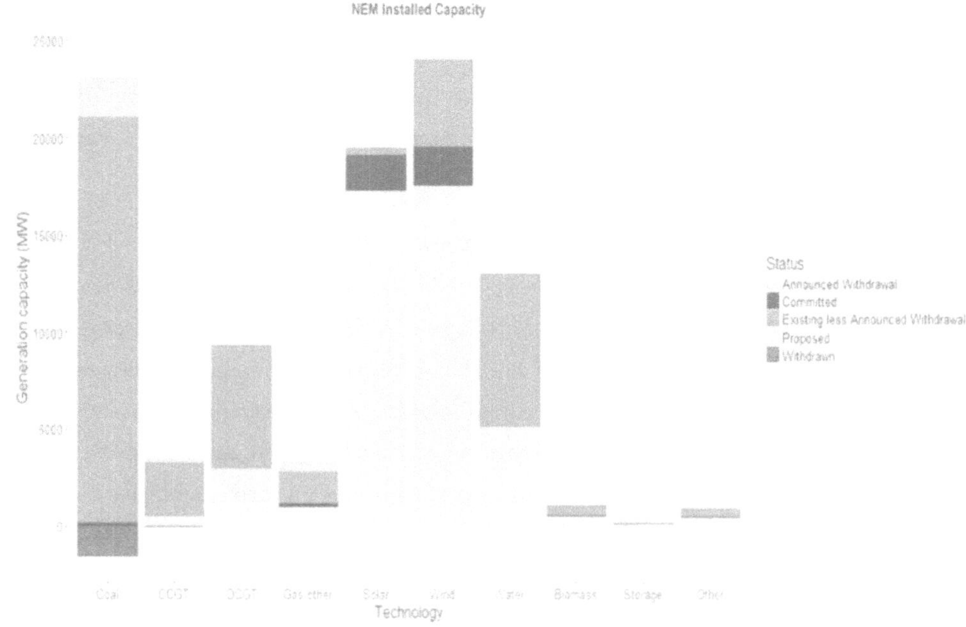

Figure 1-1: Distribution of electricity generation by technology in the NEM (AEMO, 2018b)

Energy policy and the regulatory environment in Australia's electricity sector is complex as shown in Figure 1-2 (Byrnes *et al.*, 2013). The market is de-regulated and consists of electricity generators which provide the supply of electricity, the NSPs which are either government or privately owned are responsible for the distribution of electricity. The NEM serves as a wholesale electricity market providing a market for energy to be supplied to 90 percent of Australia's population encompassing six of its eight states and territories (MacGill *et al.*, 2006). The electricity generated is sold wholesale to the retailers on the NEM at the spot price, who in turn sell on the electricity to the end consumers. Renewable energy projects commonly enter into long-term PPAs for the sale of electricity at a fixed price (Kristinsdottir, 2012). Retailers enter into fixed-price PPAs with independent generators in order to hedge against price volatility in the electricity wholesale market (Buckman *et al.*, 2014).

Other key stakeholders that need to be considered include;

AEMO, who is responsible for the wholesale and retail of electricity and gas across Australia, aside from Western Australia and the Northern Territory.   The Australian Energy Market Commission is responsible for the market development of the national electricity rules imposed on market participant. The Australian Energy Regulator has the responsibility of regulating wholesale generation, distribution and transmission of electricity in the NEM (Byrnes *et al.*, 2013).

The community is another important collective stakeholder. Social acceptance of renewables, in particular wind energy, is a factor that can often be a barrier to ambitious government targets for development. A study was carried investigating the large scale wind utility at King Island located in Tasmania, Australia. The proposed wind farm did not proceed and significant social conflict was experienced, as is often the case on large scale wind developments where localised opposition is common (Colvin *et al.*, 2016). Community engagement is something that should not be taken for granted to achieve community acceptance. There is also socio-political and market acceptance that should be considered, social-political acceptance includes acceptance on the broadest, most general level. Market acceptance deals with acceptance of the technology and method for generating and procurement of energy in the market place (Wustenhagen *et al.*, 2007).

The project financiers also play a critical role in providing project finance for the development of capital intensive projects, and is an important stakeholder group to consider. Project finance is the financing strategy through which most wind farms have been developed historically. The securing of project finance depends on a number of influencing factors. Barriers to obtaining finance in Australia in the past have been identified, and include factors such as regulatory risk surrounding legislation of the renewable energy target, semi-privatisation of retailers and capital availability following the global financial crisis (Kann, 2009). Similarly in the UK project finance for large scale RE projects slowed in response to global financial crisis. Following this in 2014 other European banks halted their commitment to signing up to long term (15 year) loans for RE projects in response to European regulation changes (Basel III requirements). Developers note that Japanese banks like BTMU and SMBC, and French banks, like Société Générale, have recently become engaged in project finance deals supporting utility scale RE development, such as wind and solar. In Australia the Clean Energy Finance Corporation is a state initiative to support the development of large scale RE, by mobilising private sector investment (Geddes *et al.*, 2018).

The purchaser of electricity, often a retailer, who purchases the electricity generated by the renewable energy plant is a critical component of the project financing structure. Both equity and debt providers require revenue certainty before being able to commit funds to a project. A PPA is agreed between the wind farm special purpose vehicle and the purchaser for the sale of generation for a certain period. This contractual arrangement provides the necessary revenue forecasting required for project financing (Kann, 2009).

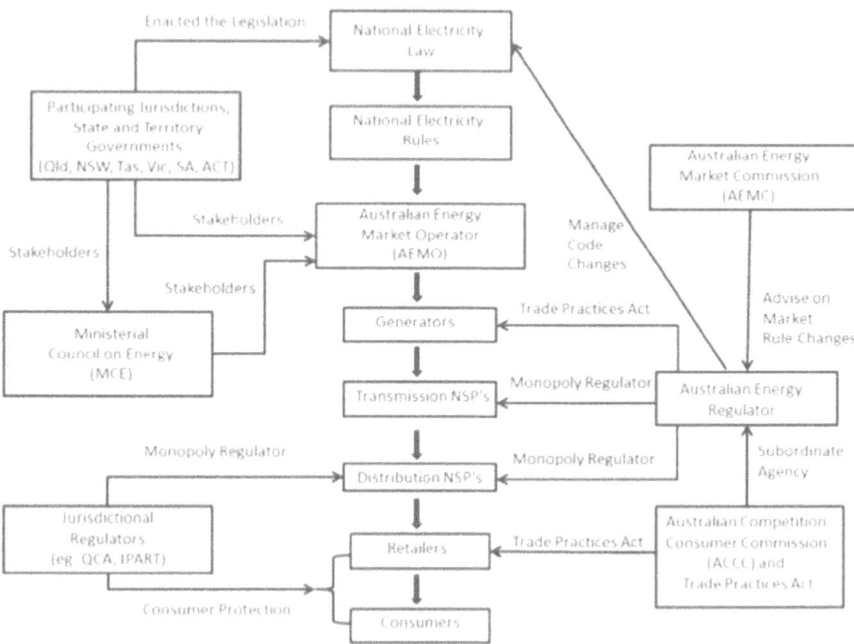

Figure 1-2: Australia's national electricity market regulatory environment (Byrnes et al., 2013: 712)

1.2.3   Barriers to RE Integration

General barriers to the adoption of renewable energy depends on various country specific factors, some of the barriers to the integration of RE generation are identified below. Specific reference to the Australian market is included where this has been identified.

The World Bank has published a Renewable Energy Toolkit (WorldBank, 2008), which outlines the various barriers to renewable energy development and an overview of the approaches to overcome such barriers. This serves as a good starting point to define the categories of barriers that may exist, before defining project specific barriers. The categories are shown below.

COST

Conventional sources of power generation are subsidised, with estimated subsidies (including the indirect environmental subsidies) at AUD $5.3 trillion in 2015 A study on China showed renewable energy subsidies lagging far behind the subsidies provided for fossil fuels (excluding the indirect costs associated with environmental impacts), which inhibited renewable energy production and investment (Ouyang and Lin, 2014). Cost competitiveness and government support for existing conventional electricity sources is noted as being a barrier to renewables development in Australia in the past (Byrnes *et al.*, 2013). Cost can be a major hurdle for renewable energy due to high up front capital costs compared to traditional generators (Byrnes *et al.*, 2013; Arnold and Yildiz, 2015). Investment in renewable energy projects traditionally required financial incentives or subsidies because such projects have higher capital costs, do not benefit from the same economies of scale as traditional generation and are considered riskier due to technological and resource uncertainties (Abdmoulehn *et al.*, 2015). It should be noted that although wind and solar PV have previously incurred higher up front capital costs per MW installed, the variable costs during operation are negligible compared to the costs of the fuel source to power conventional thermal generation plants. This is represented by the levelised cost of electricity (LCOE) that discounts the future operation and maintenance (O&M) costs of the plant to net present value (NPV), per unit of generation. The LCOE differs depending on the weighted average cost of capital (WACC) assumed, as can be seen in Figure 1-3. It should be noted that the below graph is also dependent on assumptions of capital and O&M costs as well as fuel costs, and may shift should these assumptions differ. Wind is shown as having a higher LCOE at higher WACC levels, due to its higher upfront capital costs. Wind power is shown to be the least-cost power source for WACC below eight percent (Hirth and Steckel, 2016).

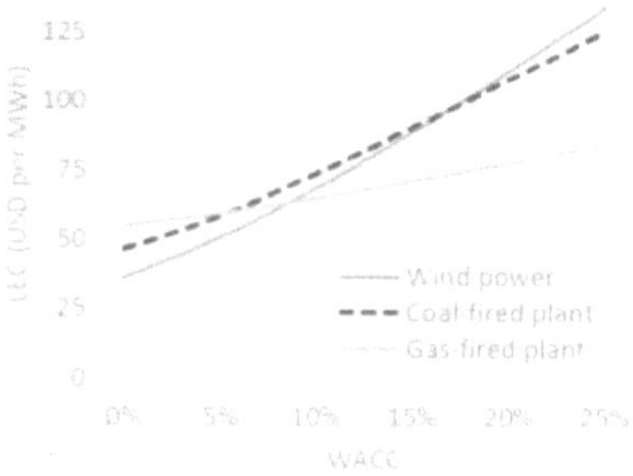

Figure 1-3: Levelised cost of electricity for different generation sources at varying WACC values. (Hirth and Steckel, 2016: 3)

LEGAL AND REGULATORY

Renewable energy projects are generally developed by Independent Power Producers (IPPs). In order for the projects to be built a supportive legal and regulatory framework is required to allow for smooth integration and development of RE (Abdmoulehn *et al.*, 2015). In countries where the electricity industry is not de-regulated, IPPs may not be able to legally connect and sell power under a PPA, unless specifically mandated by the state. The lack of certainty surrounding the legal framework creates barriers to investing in projects (WorldBank, 2008). Policy instability with sudden changes to legislation and policy being put on hold is noted as a previous barrier to renewables development in Australia (Byrnes *et al.*, 2013).

Access for projects connecting to transmission networks may also be constrained by high prices to connect or may impose unreasonable generating standards on the project. The lack of standard conditions for connection result in uncertainty in grid connection and risk to the project (WorldBank, 2008; Abdmoulehn *et al.*, 2015). Grid connection costs have also been shown to be a previous barrier for project developers of renewables in Australia, which may incorporate deeper network costs required to strengthen the network to allow for the project to connect (Byrnes *et al.*, 2013).

Obtaining planning approval can also be problematic, especially for wind projects (Wustenhagen *et al.*, 2007; Martin and Rice, 2012). State planning authorities may not be well equipped to handle applications for renewables, due to inexperience in approving similar projects and a lack of standards for siting wind turbines. Impact on migratory bird routes, visual and noise impacts and various other environmental factors need to be considered by the state authorities (WorldBank, 2008). Although some states such as

South Australia may be familiar with the approval process for RE plants, obtaining permitting may be a lengthy and complicated process due to the myriad of acts and policies (i.e. Large-scale RE Target (LRET) policy for large-scale RE plants, Biodiversity Conservation (EPBC) Act 1999) which RE developers are required to adhere to (Martin and Rice, 2015).

Access to debt and equity is considered to be one of the significant barriers to renewables development. The requirements for securing finance are linked to the previous discussions and include factors such as bankable power purchase agreements being in place (Kann, 2009), credit worthiness of off-takers, the perceived risk with 'new' technology resulting in more stringent terms being imposed (WorldBank, 2008).

A study carried out investigating the barriers to renewables integration in Queensland, Australia in 2011, found that the most significant barrier to renewable energy growth in Queensland was finance related. Firms and stakeholders confirmed that the high capital costs of large scale renewable energy projects in remote regions had an impact on investments (Martin and Rice, 2012). Despite this, it appears that development has improved significantly in 2018 with approximately 700 MW of utility scale solar installed, according to AEMO generation data published (AEMO, 2019). This indicates a potential shift in the sentiments expressed in 2011, with regards to the challenges in obtaining finance. Other finance related issues noted included finance availability, securing long term PPAs, and insufficient financial incentives.

1.2.4   Supportive Mechanisms/Frameworks

An outline of the measures introduced to address the barriers to the integration of RE is discussed, with specific reference to the Australian market where relevant.

COST

In response to the up-front cost of RE there are various economic measures that can be employed to overcome such barriers. Energy subsidies can be implemented supporting RE development. A full economic valuation of RE can help support the implementation of such incentives. The valuation focuses on aspects other than the up-front costs of RE, and takes into account price security and stability during operations, health and environmental benefits (WorldBank, 2008; Ouyang and Lin, 2014).

Financial incentives in the form of tax breaks to support utility scale RE asset investments was one of the significant items raised in response to a study investigating the actions required for the growth of the RE industry in Queensland, Australia. Issues with the implementation of tax subsidies at a state level involve the amendment of taxation law which require implementation at a federal level (Martin and Rice, 2012). Tax credit appears to be effective as it may equal a subsidy on the investment of about 25– 35 percent, depending on the profit and fiscal status of the company (Abdmoulehn *et al.*, 2015). Various tax incentive policies exist which include (Abdmoulehn *et al.*, 2015); Import duty concession which reduces the cost of importing equipment; VAT concessions which may relate to the sale of electricity being exempt from VAT or VAT rebates for project related expenses; Tax credit/accelerated depreciation which refers to the depreciation of an asset resulting in higher initial tax rebates for expenses incurred; and tax holiday allowed on the company income tax that would be payable for the companies generation income.

The Queensland government was providing over AUD $200 million in grants and funding for RE projects previously, although it was noted by study participants that this would have to be increased to meet the Queensland RE objectives (Martin and Rice, 2012).

LEGAL AND REGULATORY

The traditional approach to electricity generation and distribution is that of a state or privately owned regulated utility that owns all generation, transmission and distribution assets. This structure requires certain policy and legal frameworks to be in place, should an IPP wish to connect into the utilities network, deregulation of the electricity industry allows RE producers access to the grid (Painuly, 2001). Deregulation of the electricity sector can enable the RE IPPs to enter the market through a competitive process of generating and selling of electricity on an open market such as the NEM in Australia. Disadvantages of this approach is that generators move away from a long term PPA approach, selling generation on the spot market instead. The ability for IPPs to secure financing is compromised, which generally requires PPAs to be in place for revenue security (WorldBank, 2008; Kann, 2009).

There are various mechanisms that can be adopted for promoting the secure power purchase arrangements required for a RE project. These include feed in tariff (FIT) arrangements; tendering arrangements which calls upon RE projects to supply a fixed amount of generation to be sold under fixed term PPAs; green pricing schemes, and green certificates. Two main mechanisms used by government include FIT and tendering arrangements (Abdmoulehn *et al.*, 2015). There is no federal FIT in Australia, FIT mechanisms vary between states and territories in Australia, with most FITs being reduced or phased out (Byrnes *et al.*, 2013). A tender process is a method employed by government to set a target of installed RE capacity and award projects based on the lowest required government payment per MWh (Aparicio *et al.*, 2012). An example of this is a reverse auction scheme which involved the tendering of multiple projects competing on price in the Australian Capital Territory for the provision of 40 MW of large-scale solar capacity that operated from January 2012 to August 2013. The reverse auction demonstrated a process whereby the competitive pricing resulted in lower than expected FIT prices and at the same time providing investor certainty. The reverse auction tendering process could be an effective model for stimulating the future RE growth (Buckman *et al.*, 2014).

A lack of reasonable interconnection standards can be another barrier to RE projects. Appropriate standards are required to ensure fair pricing across all projects. Transparent standards with reasonable methods to determine access charges are required to enable cost estimations (WorldBank, 2008; Abdmoulehn *et al.*, 2015). In order to achieve the highest level of RE development, it is important to address issues surrounding grid access and to establish a transparent and economically fair charging system (Abdmoulehn *et al.*, 2015).

RENEWABLE ENERGY ENABLING POLICY

Once the legal and regulatory barriers are addressed the growth of RE can be promoted through various supportive policy being instituted. The supportive policies include the following (WorldBank, 2008);

- National targets

National targets are aimed at promoting RE development and reaching a certain MW capacity of RE by a certain date. Australia has a Renewable Energy Target (the 'RET') that was introduced in 2001. The RET requires that 20 percent of electricity generated to be sourced from renewable sources. Linked to the RET is a support mechanism that provides credits to generators per MWh produced which equate to tradeable certificates being earned (Byrnes *et al.*, 2013). In June 2016, the Victorian Government of Australia committed to a state target. The Victorian renewable energy generation target is for 25 percent of generation from RE sources by 2020 and 40 percent by 2025. The Victorian Renewable

Energy Auction Scheme (VREAS) was introduced to support achievement of the Victorian Renewable Energy Targets (VRET). A reverse auction tendering process was held in 2017, with the Victorian government calling for bids from RE developers. The Victorian government plan to support up to 650 MW of new RE generation by signing a support agreement which provides the necessary long term financial security for RE projects (VicGov, 2018a).

- Mandated market policies

Mandated market policies refer to the policies implemented by government such as feed in tariffs (an example of price setting) which allow for generators to export electricity to the market at a fixed price per unit of electricity generated (Lesser and Su, 2008).

Quantity setting refers to the implementation of requirements for electricity market participants to source a percentage of their generation from RE sources (WorldBank, 2008; Kann, 2009). An example of this is the RET regulations that requires all Australian retailers and wholesale electricity customers to source an increasing quantity of energy from RE sources. The RET announced in 2009 raised the requirements to 20 percent by 2020 (Aparicio *et al.*, 2012; Martin and Rice, 2012).

Another example of mandated market policy includes competitive tendering through a reverse auction process whereby a set quantity of RE is purchased (Buckman *et al.*, 2014). The VRET as discussed under national targets is an example of this.

- Financial/fiscal incentive policies

These are policies designed to provide financial and fiscal incentives to renewable energy producers. The policies cover various objectives which can include reducing up front capital costs (through grants), making available project finance, reducing capital and operating costs (through tax credits) and enhancing revenue streams through carbon credits. Financial incentives include public and private sector funding, public funding is provided by government through grants, low interest loans or loan guarantees. Private sector funding is provided by banks and other financial institutions which can provide favourable loans. Fiscal incentives can include tax exemptions for RE investors, as well as imposing carbon or energy taxes on fossil fuel sources (Abdmoulehn *et al.*, 2015).

The future of renewable energy development in Australia has prompted the following problem, question and proposition;

1.3 PROBLEM STATEMENT

The problem to be addressed is defined as follows:

The continued growth of RE development in Australia is uncertain, given fluctuating investment in the past and the lack of Federal RE policy going forward.

1.4 RESEARCH QUESTION

The research question to be answered is as follows:

What are the potential barriers that may need to be addressed and strategies or policies that may be required for continued growth of RE in Australia, based on the perspective of energy specialists at a leading consultancy operating in the built environment in Australia?

1.5 RESEARCH PROPOSITION

The research proposition being assessed is as follows:

Based on from the perspective of energy specialists at a leading consultancy operating in the built environment in Australia, the growth of RE in Australia is likely to be contingent upon addressing potential barriers that may emerge and introducing certain policies to incentivise investment.

1.6 AIM

The aim of the study is as follows:

To identify, from the perspective of energy specialists at a leading consultancy operating in the built environment in Australia, potential barriers that may hinder RE growth and possible strategies and policy support mechanisms that may be required for the growth of RE in Australia.

1.7 OBJECTIVES

The research objectives to be achieved are to:

Understand, from the perspective of energy specialists at a leading consultancy operating in the built environment in Australia;

- Whether RE development in Australia requires support/market intervention for continued growth.
- What the potential barriers are that may hinder continued growth of RE in Australia.
- What the strategies and/or policy support mechanisms are that could be implemented to overcome these barriers.

1.8 METHODOLOGY

Policy measures implemented in various countries to support RE have had a marked impact on its development. Different approaches have been noted in various countries for supporting RE development (Aparicio *et al.*, 2012). The policies can be structured to provide financial (grants), fiscal (taxation) and legislative (FIT or tendering) support such as tax incentives, tendering arrangements, and green certificates (Abdmoulehn *et al.*, 2015). A qualitative research methodology has been implemented in order to reveal the potential barriers and what the support mechanisms/strategies may be, which may impact the growth of RE in Australia. Qualitative research is useful in studying policy setting and the complex social and bureaucratic processes associated with the implementation of such law and policy (Denzin and Lincoln, 2011).

The methodology implemented for achieving the aims and objectives of the study is as follows:

- Carry out a literature review of the aspects inhibiting RE development and frameworks and support structures that aid in RE development.
- Review the literature on the current state of RE in Australia with regards to the framework and support structures in place as well as barriers to RE development.
- The above steps formed the basis of the research to be carried out. Formal or standard open-ended interviews were undertaken with industry participants to understand the relevance of each item identified in the literature review as well as other factors not identified in the literature.
- The findings of the research has been presented.
- The findings have been analysed by applying a pattern matching technique as the method of analysis.
- The analysis of the findings and recommendations to be presented has been concluded on.

- The implementation of RE in Australia is in a relatively early phase. Lessons learned and actual data on policy and support framework introduced is limited.
- The study is limited to the barriers or identified support mechanisms in the literature, which may not apply to the Australian context. The identification of barriers and suitability of support mechanisms is based on consultation with a limited number of industry participants and interpretation of the literature in the context of Australia.
- The efficacy of the proposed solutions is subject to evaluation by selected industry participants whose recommendations may contain biases and is limited to the extent of their knowledge.
- Participants may influence each other in their discussions and contact with similar stakeholders in the industry leading to a possible situation of group think where opinions and theories are shared.
- The scope of the study is limited to a review of the various energy policies that may be necessary to support the RE industry under an Australian context. The comparison of the efficacy of each policy is not specifically included.
- The study is based on current market conditions, such as market demand, levelised cost of electricity for RE technologies, and existing bankable technology available to the market, which may be subject to change in future.
- It is acknowledged that there may be various combinations of key factors which could be attributed to varying levels of growth in RE in Australia. The study envisages stable growth at slightly lower levels compared to 2017-2019 rates, when referring to continued development.

The research report has been structured as follows:

Chapter 2 covers the review of the literature on barriers to the implementation of RE and proposed support mechanisms. The review focuses on the status quo of RE barriers and support structures in Australia. The formal or standard open-ended interview questions for the collection of empirical data has been formulated based on the critical review of the above literature.

Chapter 3 discusses the preferred methodology selected to undertake the research and the basis for selection. A qualitative research approach was selected using a case study method. Interviews have been used for data collection. The data has then been analysed by employing a method of pattern matching.

Chapter 4 presents the data from the research and the findings identified based on the patterns that emerge from the data. The findings have been analysed with reference to the literature presented in Chapter 2.

Chapter 5 addresses the research question, aim and objective. The research proposition, being that the growth of RE in Australia is likely to be contingent upon addressing potential barriers that may emerge and introducing certain policies to incentivise investment, is shown to be supported. Recommendations and suggestion for further study have been put forward.

2.1   INTRODUCTION

The literature on energy policy is explored in this chapter, in particular such policy that provides a supportive framework for renewable energy development. The policy is analysed in terms of how it addresses the barriers to renewable energy development as well as potential shortcomings. The status of the Australian market is included as part of the analysis.

2.2   GENERAL POLICY OVERVIEW

The superiority of one policy mechanism over another is not clear when reviewing the literature. Most countries adopt a range of policies as part of a combined approach to stimulate RE development (Abdmouleh *et al.*, 2015). The optimal policy mix is also not something that can be drawn from the literature (Polzin *et al.*, 2015). The success of the policy is determined by how such policies interact, and are also influenced by other political, social and economic factors (Abdmouleh *et al.*, 2015). There is therefore no consensus as to what the optimal policy/policy combination is (Simshauser and Tiernan, 2018). The advantages and disadvantages of various policies can be discussed though, and have been outlined under the relevant sections in this Chapter. The outcome of a policy/policy mix may be different for different countries, the implementation of the same policy in different countries is emergent, displaying different policy effects (Sun and Nie, 2015).

Aquila *et al.* (2017) confirms that most governments adopt a policy mix, stating that most government policy includes a combination of short and long term policy strategy. Short term policy, which aims to support the long term policy, and are seen as supplementary and includes direct subsidies (grants), tax concessions and carbon tax. This is in contrast with Finland and Malta which have these short-term policies (tax incentives and investment grants) as their major policy prior to 2011 (Klessmann *et al.*, 2011). The most significant long term policies that have been implemented to support RE development include; feed-in tariff (FIT), auction schemes (or tendering schemes) and quota systems (or renewable portfolio standard (RPS)) (Aquila *et al.*, 2017). Sun and Nie (2015) confirms feed-in tariff and RPS to be the dominant policies in place, with 60 countries worldwide adopting one of the two policies. RPS is excluded from this statement as a dominant policy measure. Abdmouleh *et al.* (2015)confirms FIT and auction schemes as dominant policy measures, with reference to RPS as an emerging policy (implemented along with green pricing certificates). Klessman *et al.* (2011) states that FITs are the dominant policy in the European Union, with six countries adopting RPS, whilst auction schemes were no longer used as the main policy scheme. Australia's Renewable Energy Target (RET) which is an RPS scheme was introduced in 2001 (Simpson and Clifton, 2014b). The RET was increased to 20 percent of generation to be sourced from renewable sources by 2020. The obligation was reduced from 41,000 GWh to 33,000 GWh in 2015, with the Turnbull administration confirming that the scheme would not be extended beyond 2020 (Diaz-Rainey and Sise, 2018). Australian State governments have also introduced their own renewables targets, with the objective of implementing through State auction schemes in Victoria, to achieve 40 percent penetration of RE by 2025 (VicGov, 2018a) and Queensland, to achieve 50 percent penetration of RE by 2030 (QueenslandGovernment, 2018).

Although the optimal policy mix is debatable, researchers agree on the importance of long-term policy commitment in encouraging RE development (Polzin *et al.*, 2015). The availability of affordable finance is seen as one of the major barriers to entry for RE developments (Eleftheriadis and Anagnostopoulou, 2015). Long term policy commitment which aids in stability and predictability increases confidence for investors, through the reduction of regulatory risk, which lowers financing costs (Klessmann *et al.*, 2013).

2.3   FISCAL ASPECTS

Fiscal incentives are considered to be a compliment to major support policy such as a FIT or an auction scheme. Fiscal intervention can be in the form of government issued grants and loans, tax breaks and introduction of a carbon tax (Abdmouleh *et al.*, 2015). These fiscal policies are discussed in more detail in this section.

2.3.1   Grants and Loans

Public sector funding can be used to support the development of RE projects, through the application of grants or loans. Grants offer financial assistance to eligible recipients which do not have to be repaid. Low interest loans can be issued by national or regional financial institutions to eligible recipients. Unlike grants these are required to be re-paid (Abdmouleh *et al.*, 2015). Kitzing *et al* (2012) states that most European countries have implemented some sort of investment grant scheme for RE. The grants range from 5 percent to more than 70 percent of the total investment cost (Kitzing *et al.*, 2012).

Grants and loans have been shown to be an effective measure in relieving short term financial constraints, by reducing the cost of finance for a project (Polzin *et al.*, 2015). Grants and loans depend directly on public budget, which in turn makes them more unstable and a less reliable mechanism to spur on the industry (Johnstone *et al.*, 2010).

2.3.2   Tax

Fiscal support mechanisms in the form of tax relief can be implemented in various ways. These include, income tax relief either directly or indirectly through mechanisms such as accelerated depreciation of assets. Reduced Value Added Tax (VAT) or Good Service Tax (GST) in Australia can be applied to sales on the assets generation. Reduction of import taxes is another strategy which can be beneficial to most RE projects, which require a significant share of the components to be sourced from abroad (Aquila *et al.*, 2017). Tax exemptions can encourage companies to invest in RE projects, tax credits can effectively translate into a subsidy for a RE development (Abdmouleh *et al.*, 2015). Although this approach rewards investment in RE (Abdmouleh *et al.*, 2015), the main shortcoming of providing tax incentives is that it is dependent on public budget and therefore susceptible to fluctuation depending on government support (Polzin *et al.*, 2015).

Environmental taxes such as a carbon tax can be levied against carbon producing generating plant, based on the quantity of carbon dioxide expelled during the generation of electricity. The revenue from the tax can then be re-distributed to RE generators (Abdmouleh *et al.*, 2015) to spur development. Carbon tax is one of the most popular fiscal mechanisms for internalising the cost of carbon emissions (Aquila *et al.*, 2017). The purpose is to level the playing fields between conventional and RE generators, with the former being heavily subsidised in the past (Polzin *et al.*, 2015). The internalisation of carbon dioxide emissions (carbon pricing) through emissions trading occurs in many countries. The US states of Connecticut, Delaware, Maine, Maryland, Massachusetts, New Hampshire, New York, Rhode Island and Vermont operate carbon pricing schemes. China launched its emissions trading scheme at the end of 2017 (Simshauser and Tiernan, 2018). The tax imposed on conventional generators is ultimately passed on to the consumer though, impacting on electricity prices especially in a system which is heavily dependent on conventional power generation. Australia implemented a carbon tax in 2012 with an intent to transition to an emissions trading scheme, the legislation was however repealed in 2014 due to partisan differences (Diaz-Rainey and Sise, 2018). The tax was seen by conservative members as a threat to low cost conventional power generation (Simshauser and Tiernan, 2018).

2.4.1   Feed in Tariff

The objective of a FIT system is to enable a RE generator to sell its generation at a fixed price for a specific time period. The tariff is set by the authority, based on the generation type (Fouquet, 2013). The long term purchase agreement for the sale of electricity is focussed on providing support for developing new RE projects (Abolhosseini and Heshmati, 2014). Important aspects of a FIT policy include provisions for access to the grid, a long term power purchase agreement, and prices that reflect the levelised cost of electricity for the generator (Abolhosseini and Heshmati, 2014). The above is echoed by Aquila *et al.* (2017) in that a FIT programme should incorporate aspects such as guaranteed access and priority connection to the grid. This is enabled through clear and transparent grid connection rules and regulations. The priority for connection should assist in reducing timeframes associated with obtaining a grid connection. Long term purchase agreements with guaranteed revenues avoid any exposure of the project to sale of generation and fluctuation of pricing (Aquila *et al.*, 2017).

The literature indicates that the FIT approach is the dominant scheme implemented for the development of RE. FIT schemes have been shown to be appropriate in developing countries with the goal of expanding their RE portfolio (Aquila *et al.*, 2017). Amongst the major policy mechanisms, FIT schemes are clearly dominant (Kitzing *et al.*, 2012). Polzin *et al.* (2015) similarly highlights that various sources in the literature underline the effectiveness of FITs to spur deployment and lower the otherwise inherent risk in investing in RE developments (Polzin *et al.*, 2015). Aquila *et al.* (2017) similarly notes that FITs present higher efficiency in promoting RE developments due to the risks that a FIT scheme mitigates against. Sun and Nie (2015) also note that several researchers have concluded that FIT was the most widely adopted policy to stimulate the rapid development of RE. They concluded that compared with other policy mechanisms, FIT can deliver RE more effectively and at lower cost (Sun and Nie, 2015). Polzin *et al.* (2015) agrees on the superiority of FITs to spur RE development and to lower the associated risks. However it is noted how FITs proved particularly successful in countries such as Germany and Italy with some exceptions in other countries such as Spain (Polzin *et al.*, 2015). Abolhosseini and Heshmati (2014) also note the significance of FITs in being effective in the promotion of the deployment of RE in Europe.

However, they do raise shortcomings of the FIT schemes in creating a single electricity market across Europe and conclude that no single policy is capable of solving the issue of RE development in a consolidated European market (Abolhosseini and Heshmati, 2014). Schaffer and Bernauer (2014) acknowledge that most observers agree that FIT systems have had a major impact on renewable energy production, but note that it is challenging to control costs of a FIT system to end consumers, and to set tariffs that reflect the levelised cost of electricity (Schaffer and Bernauer, 2014).

2.4.2 Renewable Energy Auction

The objective of renewable energy auctions is to reduce prices through a competitive bidding process. This is considered to be the most favourable policy for the reduction in end sale price of electricity, due to the reduction in price due to market based pricing (Abdmouleh *et al.*, 2015).

Tenders are usually implemented in conjunction with a complementary policy. The policy combination can be adjusted to suit the specific characteristics of the market. In an auction, the facilitator would put out a tender document detailing the requirements of the auction scheme. The tenderers would then bid against each other to fulfil the requirements of the tender, outlining their specific level of support/commitment (LGC price, tariff, local content etc.) depending on the characteristics of the auction. The most attractive tenders are then awarded the bid (Kitzing *et al.*, 2012).

Although auctions are considered to be advantageous in their ability to drive down price of generation, and promote other requirements, such as local content requirements, extremely low cost bids has shown to negatively impact investor appetite, with smaller margins leading to greater risk, leading to lower levels of investment (Abdmouleh *et al.*, 2015). Aquila *et al.* (2017) confirms this view that bids with excessively low pricing, that do not reflect the technology cost sends an unfavourable message to the industry, such as in Brazil in 2012, which can lead to low project realisation rates (Aquila *et al.*, 2017).

Kitzing (2016) provides specific reference to industry examples which exhibit a significant cost reduction due to the implementation of an auction scheme; tariffs in California fell from €79.5 / MWh in the first auction round to €70.5 / MWh in the third round. Brazil showed reductions in the order of 60 percent on the implementation of an auction scheme, compared to the previous FIT policy. South Africa's auction scheme has seen increased levels of competition driving down tariffs in subsequent rounds, from R3.27 in the first round to R0.77 in the fourth round. In Germany, prices fell from €91.7 / MWh to €72.3 / MWh from the first to subsequent rounds (Kitzing, 2016). Aquila *et al.* (2017) reiterates Kitzing (2016) and Abdmouleh *et al.* (2015) position in that the disadvantage of the competitive bidding is that emergent tariffs can be unreasonably low, as a result of underbidding. The leads to low realisation rates of projects as well as delays, which have been shown in case studies of projects awarded under tender schemes (Kitzing, 2016). Spain held an auction for onshore wind and biomass in 2016. The market climate was that of high competition due to many projects that had been previously put on hold, ready for bidding, with developers under pressure to secure tenders. This led to aggressive bidding with the majority of bids unreasonably low, with doubt that some projects will be able to be realised in 2020 (Del Rio, 2016). Conversely, certain scenarios can emerge in which the developers are paid in excess of technology costs leading to windfall profits, and higher than required tariffs under a FIT scheme (del Río and Linares, 2014).

The market should therefore be evaluated prior to engaging market participants in an auction, which will be sensitive to a number of factors (Kitzing, 2016). Factors which can impact on an auction scheme include, and potential mitigation measures include; a low number of participants leading to higher tariffs and conversely, high competition leading to underbidding. An auction schedule can assist in negating late/desperate bidding. Pre-qualification and penalties can be imposed on tenderers in order to avoid speculation leading to low project realisation rates. This should be balanced so that competition is not adversely affected; single bidders may dominate the market leading to lack of competition, this can be avoided by implementing maximum bid constraints or imposing a minimum number of bidders requirement. Ceiling prices can be implemented as a mechanism to control policy expenditure. It also assists in mitigating excessive remuneration to participants. Introducing a ceiling price may have secondary effects. It can act as a target in a climate of low competition, negating the purpose of an auction. If it is set too low, it may send a negative signal to investment leading to decreased RE deployment (Kitzing, 2016).

2.4.3   Renewable Portfolio Standard (RPS)

In a quota system with tradable green certificates, also known as a Renewable Portfolio Standard (RPS), liable entities (i.e. energy suppliers) are required to maintain a specified percentage of RE certificates, generated from RE sources (Abdmouleh *et al.*, 2015). This is imposed by the quota on liable entities, as to what the percentage share of generation supplied should be renewable. Entities can purchase green certificates generated by RE plants which equate to capacity generated (i.e. 1 certificate may represent 1 MWh). The green certificates are freely traded at the market price. The price of green certificates and remuneration for renewable energy generators producing such certificates is dictated by supply and demand. The market is in theory, self-regulating, when excess certificates are generated than is required under the set quota, the price will decrease due to excess supply, reducing this additional revenue stream for generators and decreasing the incentive for future investment in new generation (Fouquet, 2013). This is in contrast with the FIT scheme which guarantees a set tariff over time, determined by policy (Kitzing *et al.*, 2012). The RPS scheme relies on the market for sale of electricity and purchase of green certificates for an additional revenue source (Aquila *et al.*, 2017). The RPS approach is not as mature as FIT and auction schemes, however is gaining popularity due to the limited government involvement, with the self-regulating nature of the market in terms of supply and demand for green certificates (Sun and Nie, 2015). It is noted by Fouquet and Johansson (2008) that the major support mechanism in EU member states are FIT and RPS schemes, indicating that the RPS scheme is actually fairly mature.

RPS schemes are usually implemented as a sole policy mechanism, although has also been seen to be incorporated with FIT, indicating a move to a balance between a pure quantity controlled market based mechanism, to quantity and price control under FIT (Kitzing *et al.*, 2012). In theory, under perfect market conditions the RPS scheme is viewed as the most efficient way of delivering a set quota or target for renewable energy. In practice, the PRS scheme has failed to achieve in some cases on delivering the set quotas (Schaffer and Bernauer, 2014). Investors face higher risks than conventional FIT schemes, as they are exposed to market price fluctuation of certificate prices and electricity prices. The RPS scheme was initially favoured by the EU commission, but faced major opposition from member states, preventing it from becoming the prevailing policy mechanism in Europe (Schaffer and Bernauer, 2014).

Abdmouleh *et al.* (2015) echoes the above sentiment, that in practice, RPS schemes fall short in delivering compared to conventional FIT schemes, which have been shown to be more effective and resulted in lower electricity prices than RPS schemes. This is supported by evidence from Europe which shows countries that were most successful in increasing their share of RE (Germany, Spain and Denmark) implemented FIT schemes. UK and Italy which implemented RPS schemes have been less successful (Abdmouleh *et al.*, 2015). Polzin *et al.* (2015) notes that RPS schemes in the US have shown mixed results, although overall have been shown to be inferior to FIT. Investors are not necessarily opposed to the market based policy mechanism, but require comfort in that the policy is coupled with reliable support for their investments (Polzin *et al.*, 2015). The context in which the RPS scheme is implemented should be taken into account, in developing countries for instance, the scheme is not considered to be favourable, due to the higher uncertainties and market risk associated with the scheme (Aquila *et al.*, 2017).The RPS scheme is considered to be a least cost policy option for governments (Abdmouleh *et al.*, 2015), which has its own advantages, despite the FIT being widely adopted for promotion of RE.

2.4.4   Grid Access

A secure and reliable grid that serves to transport electricity, as well as energy management based on demand side management is crucial to the effective deployment of RE (Fouquet, 2013). Adequate access to the grid is generally a challenge facing RE development. This is partly due to RE projects, which are generally characterised by small scale developments, often located in rural environments where grid connections are limited or unavailable (Abdmouleh *et al.*, 2015). Grid related infrastructure is costly and a lack of development of grid infrastructure has been one of the biggest impediments to the expansion of RE (Eleftheriadis and Anagnostopoulou, 2015). This is echoed by Klessmann *et al.* who states that gird related barriers are identified as major obstacles for RE projects, and concludes that administrative and grid related issues are some of the greatest barriers to RE development across Europe (Klessmann *et al.*, 2011).

There is usually a link between the levels of RE installed and the extent of a countries grid constraints. Grid congestions and the need for additional infrastructure and capacity is usually correlated with the quantity of RE already installed. Intuitively, congestion increases as RE deployment reaches higher levels (Klessmann *et al.*, 2011).

The administrative process with regards to obtaining access to the grid can also be a barrier to development. This may be due to network operators not being open to RE, lack of transparent procedures in obtaining grid access, approval times may be lengthy, and grid connection charges may be unreasonable (Klessmann *et al.*, 2011).

A summary of the grid connection issues and potential solutions has been included in Table 2-1.

Barrier	Solution
Administrative/legislative procedures (transparency, cost allocations, process and approval times, ease of access)	Guaranteed/priority grid access to RE developers
	Transparent and non-discriminatory procedures
	Regulation of grid connection cost
	Strict control of regulations by energy regulator
Grid capacity physical/technical constraints	Mandate to expand grid network and long term planning
	Compensation to RE projects during curtailment

2.4.5 Political Transparency

The overarching requirement for successful RE development is transparent and stable political support. This is a precondition for the development of RE, which requires stable political support through transparent and manageable bureaucratic procedures. Rules, such as the development, construction and operations of RE plants should be clear and concise, and should ideally be summarised under a single legislative programme (Polzin *et al.*, 2015). A long term government policy that results in stable and predictable cash flows over the life of the project is favoured by institutional investors, and the transparency of potential changes to policy is paramount in securing investment and subsequent growth in RE development (Polzin *et al.*, 2015). Abdmouleh *et al.* (2015) concurs with the above sentiment in that strong and well established political support at national, regional or local level is a key component of successful RE development. Political support and transparency refers to all aspects of energy policy, such as the regulatory frameworks and support mechanisms for the incentivisation of RE development. An important aspect of providing certainty in investment is establishing a renewable energy target that the government is committed to achieving (Abdmouleh *et al.*, 2015).

2.4.6   Technological

Investment into mature RE technologies such as onshore wind and hydro has been favoured by investors in the past. This is based on the philosophy that investment into mature technologies will result in a reduction in the cost curve much more rapidly, when compared to a revolutionary technological advancement (Masini and Menichetti, 2013). Alternative innovative technologies which may result in improved performances and greater margins on returns, generally require higher upfront capital costs, which they have failed to secure. The lack of investment into potentially superior technological innovations, is a barrier inhibiting technological advancement (Menanteau, 2000).

These RE technologies and innovations which are less mature compared to conventional forms of RE generationrequire support in order to overcome technological challenges and be fully developed and invested in at the utility scale. This support can come from government R&D initiatives, which can assist technology in overcoming technological challenges, becoming more cost competitive and demonstrating functionality at a utility scale. This can provide investor confidence and enable the technology to be employed once it has been optimised for construction and operations. Similar R&D programmes can be focussed at more mature technologies, to optimise certain processes and allow such technologies to function more efficiently and cost competitively (Abdmouleh *et al.*, 2015).

The intermittent nature of renewables are associated with the technical challenges that face grid network, especially in areas where the grid is weak. Improving the grid infrastructure is costly, and with the increasing penetration of RE into the network, further research is required into the most cost effective way to manage increased RE deployment and associated grid infrastructure. Research is focussing on changing the way to manage energy in the system through smart grid technologies and demand side management. This shift in energy management requires intervention and support by the relevant government agencies (Abdmouleh *et al.*, 2015).

2.5   AUSTRALIAN ENERGY POLICY AND STATUS QUO

2.5.1   Background to Australian Energy Crisis

Bipartisan support in Australia is expected to benefit a selection of policies in the long term that will support development. Political action is usually in the short to medium term and it is a challenge to achieve stable long term policy when governments have differing vested interests (Byrnes *et al.*, 2013).

The governance of Australia is further complicated by a dual layer of state and federal interventions. States have forged ahead, implementing their own RE policies in the absence of federal intervention, which has suffered from a lack of energy policy stability, required for investment (Diaz-Rainey and Sise, 2018). State governments are demonstrating intent in support of RE development, the Queensland and Victorian government, respectively have committed to a target of 50 percent RE by 2030 and 40 percent by 2025 (Nelson *et al.*, 2017). The Australian Capital Territory has committed to a target of 100% by 2020 (McLeod, 2017).

Australia's RE policy commitments can be traced back to 1997. The Howard Government that was the current administration at the time issued a climate change policy which was titled 'Safeguarding the Future: Australia's Response to Climate Change'. The Howard government announced that the federal government would work with state governments to achieve a target of 2 percent of electricity to be sourced by retailers from RE sources by 2010. This was the first renewable energy target actioned at national level. Mention was also made to an emissions trading scheme being implemented. RETs and carbon pricing were the two main policy measures being considered at this time (Simshauser and Tiernan, 2018). In 2011, Australia took action on steps towards an emissions trading scheme, introducing a carbon tax, with the intention of moving towards emissions trading in 2014 (Diaz-Rainey and Sise, 2018). A carbon price was established through the Clean Energy Future package but was then repealed within three years in 2014 (Nelson *et al.*, 2017). The RET incentivised by a green certificate market, introduced in 2001 was still in place though and has supported investment in RE, it is noted though that policy uncertainty around the scheme has also impacted on investor confidence (Diaz-Rainey and Sise, 2018). A range of state and federal policy has assisted in boosting RE investment over the last decade. Such policies existed to encourage the integration of low emission technology for energy generation. These policies included; The NSW Greenhouse Gas Abatement scheme; the Queensland 18 Percent Gas scheme; the RET; various energy efficiency policies and solar feed in tariffs (Nelson *et al.*, 2017).

2.5.2   Energy Policy Status Quo

In 2009 legislation was passed which set out the requirement for the 20% contribution of small and large scale RE to the total energy mix by 2020. The figure that was adopted under legislation was a total of 45,000 GWh of energy to be sourced from RE generation by 2020 (Simpson and Clifton, 2014b). In 2011, the RET was divided into two parts, namely the Large Scale Renewable Energy Target and the Small Scale

Renewable Energy Target. The LRET figure was set at 41,000 GWh to be achieved by 2020 (Prest and Soutter, 2018). Further uncertainty ensued following a legislated review of the RET (Climate Change Authority Review) through a detailed consultative process which was conducted in 2012. A number of changes were considered for the scheme and were under evaluation including the GWh target that had been set previously. A total of 34 recommendations were listed by the Climate Change Authority, with only six new recommendations being endorsed by the Commonwealth Government (Simpson and Clifton, 2014a). The Warburton Review was commissioned by the then Abbott administration in 2014. The primary recommendation of the review was to reduce the RET from 41,000 GWh to 33,000 GWh. Investment certainty plummeted following the review, evident by an approximated 88% fall in RE investment (Prest and Soutter, 2018). The RET, following the Warburton Review was subsequently decreased to a goal of achieving 33,000 GWh from RE sources by 2020 (Byrnes and Brown, 2015). This target will be fully subscribed, with projects under construction in 2018, which will enable this to be achieved once generating (Simshauser and Tiernan, 2018). Beyond this, a policy vacuum exists at a national level for achieving Australia's 2030 climate change commitments. Australia's has committed itself with bipartisan support to obligations set out under the Kyoto Protocol and the Paris Agreement. The Paris Agreement includes a commitment to reduce emissions by 26-28 percent below 2005 levels, by 2030. There is however a lack of direction with regards to energy policy at a federal level that would support these objectives. Energy infrastructure requires significant investment over the long term. The common strategy of holding onto an option and waiting for policy certainty is generally preferred with the large quantity of equity and debt finance at risk. In contrast, state governments have displayed intent on tackling climate change through implementation of various policies, such as reverse auctions, and investment in battery storage (Simshauser and Tiernan, 2018).

Policy stability is needed, in order to avoid investments drying up due to a lack of market confidence (Diaz-Rainey and Sise, 2018). This has been a challenge for national government in developing stable bipartisan policy for the long term. Following the state wide black-out in South Australia, and with the goal of examining how the NEM should adapt to the changing energy generation mix, the national government ordered a review of the NEM in October 2016. The review was to be spearheaded by Dr Alan Finkel, and was dubbed the Finkel Review, released in June 2017 after eight months of consultation. The key focus of the review was to advise on climate change policy architecture in the context of increasing intermittent RE supply. The issue to be tackled as part of the investigation was the solution to providing energy that is affordable, a supply that is reliable and sourced from low carbon emission sources. The outcome of the review was a list of 50 recommendations in order to address these issues.

The key emerging recommendation of the Review was put forward as the Clean Energy Target, which would replace the RET in 2020. Other key recommendations also included the role of gas in the move away from conventional power sources, the importance of integration and planning for the grid across the NEM and regional reliability standards, and establishment of an Energy Security Board. This initiative was never realised though, with the policy blocked by the conservative backbench. The policy was ultimately abandoned in October 2017. Following the abandonment of the Clean Energy Target, the Turnbull administration proceeded to develop a new policy out of the remnants from the previously discarded policy, which included a generator reliability obligation as one of the commitments. The revised policy did however stem from recommendations of the Finkel Review (Simshauser and Tiernan, 2018; Guidolin and Alpcan, 2019). The revised policy was dubbed the National Energy Guarantee, and aimed to solve the pricing crisis and provide policy stability in addressing each prong of energy security, affordability and emissions (Diaz-Rainey and Sise, 2018). The policy aimed to incentivise retailers to source lower carbon emission generation to achieve targets in line with the Paris Agreement. Retailers would also be required to source an amount of 'firming' generation to ensure reliability of supply (Nelson, 2018). Simshauser and Tiernan (2018) query whether the National Energy Guarantee will put an end to Australia's two decades-long RE policy instability, or be destined for disposal in the overflowing climate policy garbage bin. This appears to be the case, as at the time of writing, the current administration had rejected the National Energy Guarantee, with no policy measures suggested to replace it. It is noted that other key recommendations stemming from the Finkel Review have been accepted, particularly the need for planning at a grid wide level to overcome grid constraints and AEMO's role in the implementation of this. This has been demonstrated by the Australian Electricity Market Operator (AEMO) releasing the Integrated System Plan (ISP) in July 2018 (AEMO, 2018a), outlining the proposed future energy mix and requirements in terms of grid strengthening and transmission expansion that would be required. New projects have already been realised for strengthening of transmission networks in Queensland, to enable RE project development in north Queensland, as well as the strengthening of the NSW/SA interconnector(Stocks *et al.*, 2019).

Despite the lack of political stability at the federal level, and lack of bipartisan commitment to a single scheme, there has been an increase in investment in the RE sector since 2015, under the Turnbull administration (Diaz-Rainey and Sise, 2018). Simshauser and Tiernan (2018) attribute this to years of RET policy uncertainty which had stalled RE investment between 2013-2015. Thus, by the time that the RET was revised in 2015 there was limited time to meet the 2020 target. Green certificate prices surged due to increased obligations for retailers to purchase and limited supply, and subsequent RE project commitments soared to record levels (Simshauser and Tiernan, 2018).

2.5.3   Fiscal Aspects

GRANTS AND LOANS

Australia is one of few countries that has a dedicated green investment bank. The Clean Energy Finance Corporation (CEFC) was established in 2012 and by 2017 it had made commitments in debt financing of over AUD $4 billion, to projects with a combined value of over AUD $11 billion. CEFC's mandate is to finance clean energy development, such as building energy efficient houses for low income families, or providing financing for utility scale RE projects (Diaz-Rainey and Sise, 2018). The CEFC supports projects that are in the advanced stages of technological innovation (i.e. mature technology such as wind and solar PV). It is independent from government and run by a select group of investment managers, banking and clean energy experts. The function of the CEFC has been to remove the financing barriers that exist for suitable projects to be realised (Byrnes *et al.*, 2013). The initial budget apportioned to the CEFC was AUD $10 billion, to provide finance support over a period of five years from 2013 (Hua *et al.*, 2016). The CEFC is an important instrument that provides funding for utility scale RE development, and its potential to be relinquished is a concern for RE financing going forward (Byrnes and Brown, 2015).

In addition to the CEFC, the Australian Renewable Energy Agency (ARENA) was also established in 2012, with a remit to provide funding support to clean energy research and development and provide funding for innovative projects (Diaz-Rainey and Sise, 2018). ARENA has a budget of AUD $2.5 billion to carry out its mandate (Hua *et al.*, 2016). Although ARENA receives bipartisan support, the level and mandate for its funding is not wholly supported (Byrnes *et al.*, 2013). The national government has been restrained from being able to abolish ARENA, through the actions of the senate, and the agency continues to fund projects and has indicated that it will continue with its mandate until it is abolished (Byrnes and Brown, 2015).

2.5.4 Regulatory Aspects

The RET aims to promote the development of RE through the issuance of green certificates for RE generation. Large scale utilities are provided with certificates for their generation, one MWh of generation earns one green certificate. Liable entities (which are generally electricity retailers) are obliged to purchase a certain quantity of certificates. The system known as RPS, has been key to facilitating RE development in Australia (Byrnes and Brown, 2015). The RPS policy tool was designed in the USA, although Australia became the first country to legislate a RPS, issuing a mandate to source an additional two percent of generation from renewable sources (Cludius *et al.*, 2014). Australia's RPS became known as the mandated renewable energy target (MRET). The policy aimed to increase renewable generation share from 10-12 percent. Liability was placed on electricity retailers to purchase a tradable green certificates generated from RE sources (Byrnes *et al.*, 2013).

In 2007, following the election of the Rudd Labour government, a move was made to legislate the 20 percent RET to be achieved by 2020, which was subsequently passed in 2009 (Cludius *et al.*, 2014). There is currently no intention of extending the target beyond 2020, as such, investment incentives at the national level beyond the 2020 horizon are unclear (Diaz-Rainey and Sise, 2018). The RET contained a number of design issues, which included a two-year statutory review of the RET. The result of such reviews was a predictable investment freeze prior to scheduled reviews, at the risk of potential policy change. Investment was characterised by sharp spikes and dips during review periods (Simshauser and Tiernan, 2018).

An increase in the penetration of RE into the market puts downward pressure, in theory, on the wholesale spot price of electricity. This is as a result of the close to zero marginal cost of RE, which does not require a minimum operational cost to run with the fuel source being free. This displaces higher operating cost conventional generation (merit order effect) in the market. A higher quantity of RE is therefore procured due to the lower cost to meet demand, with the spot price of electricity reducing as a result (Cludius *et al.*, 2014).This phenomenon is also elaborated on by Blazquez *et al.* (2018) and Nelson (2018). The actual case in Australia though has been a dramatic upward shift in wholesale energy prices due to significant capacity of aging coal fired power stations being retired (Nelson, 2018). The Australian Competition and Consumer Commission (ACCC) launched an investigation into the causes behind the steep rise in wholesale prices. The investigation concluded that the main factors contributing to the increase were a tightening of supply due to the exit of coal-fired power plants; along with increased market concentration through acquisitions and closures of large generators affecting bidding behaviour; fuel source costs including a shift in generation type; and a lack of competitively priced gas (ACCC, 2018).

Two types of exemptions exist under the RET legislation, these include self-consumption and emissions intensive trade exposed industry (EITEI) (Cludius *et al.*, 2014). The exemption to emissions intensive industry increases the burden of the certificate costs to those non-exempt parties (i.e. such as households). The rationale behind the exemption is that trade exposed industry would be subject to a competitive disadvantage when passing on certificate costs, when compared to other countries which may not be subject to emissions costs. It is argued by Cludius *et al.* (2014), that there is scope for re-examining the exemptions to EITEI given the theoretical benefit of the downward pressure on wholesale prices due to increased RE penetration, that the exempt industries may benefit from. As mentioned above, the dramatic upward shift in wholesale pricing does not reflect the theoretical downward pressure of increasing low marginal cost RE penetration. As such, any exemptions to EITEI industry would only be sensible for consideration at a point in time when there is a visible impact of RE on wholesale energy prices. Nelson *et al.* (2015) confirms the theoretical notion of a reduction in wholesale prices with increased penetration of RE, noting that additional subsidy would be required in order for new RE plant to attract investment, to cover long run (i.e. sunk costs) marginal costs. This would necessitate an increase in green certificate costs or another subsidy, which is ultimately passed on to the consumer (Nelson *et al.*, 2015). The conclusion is the same in that EITEI industry may benefit in future, should market dynamics reflect a shift in energy costs being captured under green certificates as Nelson *et al.* (2015) proposes, as opposed to wholesale energy prices.

The effectiveness of RPS is debatable, with evidence suggesting that feed in tariffs may be more effective in promoting RE (Diaz-Rainey and Sise, 2018). This position is echoed by Byrnes *et al.* (2013), in that FITs that promote utility scale RE projects and that are adjusted based on the technology maturity and stage of integration, may prove to be more efficient than current policy measures. FITs may be a more practical approach to accelerating the development of RE in Australia, as they act to provide certainty around revenue returns, encouraging investment (Byrnes *et al.*, 2013). A FIT system that is reviewed and potentially adjusted on a regular basis, to ensure that windfall profits are not realised, market distortions are abated, artificial markets are not created and the status of RE integration in considered, can assist in providing investment certainty and increasing RE capacity in Australia (Byrnes *et al.*, 2013). The analysis of the state targets below, indicate that a feed in tariff structure, utilising a 'contract for difference' approach is the preferred policy tool being implemented by state governments. A contract for difference enables the generator to set a base price (set tariff) for the electricity to be sold, providing revenue certainty for investors. Should the pool price be higher than the contractual base price, the difference is paid by the generator (seller) to the buyer, and vice versa in the case of the pool price being lower than the base price (Buckman *et al.*, 2014).

STATE TARGETS

In the absence of a coherent national climate change policy, some state governments have proceeded to set their own RE targets. For instance, in 2007, South Australia set out to achieve 20 percent RE integration at state level by 2014 which was subsequently achieved by 2011. A revised goal of 33 percent RE integration by 2020 was therefore set. Victoria, similarly set a goal of 25 percent installed capacity by 2020 (Aparicio *et al.*, 2012). State targets have shown to be more ambitious than any proposed national targets or obligations under the Paris Agreement, as discussed below.

AUSTRALIAN CAPITAL TERRITORY (ACT)

The ACT implemented its own legislation in 2010, the ACT parliament legislated the Climate Change and Greenhouse Gas Reduction Act that legislated a 40 percent reduction on 1990 level of emissions by 2020, 80 percent reduction by 2050, and a complete reduction to zero percent emissions by 2060. Compared to the national targets the ACT legislation is much more ambitious (Buckman *et al.*, 2014). The scheme was based on a FIT mechanism using a contract for difference approach, which required retailers to make FIT support payments to generators for RE purchased. These costs/savings were subsequently passed on to end consumers. Proposals that were submitted under the reverse auction were required to be prequalified to mitigate against the risk of projects being bid at a low base price, which were not actually feasible. This is often seen as one of the significant risks of a reverse auction process (Buckman *et al.*, 2014). The ACT reverse auction demonstrated that by simultaneously creating investor confidence and creating a competitive environment to achieve low FIT base tariffs, RE integration could be promoted and developed effectively (Buckman *et al.*, 2014).

A study which incorporated a number of firms and stakeholders in Queensland's energy sector indicated that RE projects face challenges in Queensland such as high capital costs, insufficient financial incentives, shortages in technical workforce and high degrees of administrative red tape (Martin and Rice, 2012). This is in contrast with what is currently being promoted and acquired by the Queensland government, and indicates a shift in the momentum of RE deployment since 2012. The Queensland state government has introduced a target for RE procurement that is more ambitious than any national initiative or what is outlined under the Paris Agreement for 2030 emission levels. The Queensland government has instituted a target to achieve 50 percent renewable energy integration by 2030. The state aims to facilitate the next wave of up to 400 MW of diversified RE, which shall include 100 MW of energy storage. The mechanism for procuring this capacity is through a reverse auction process under the Powering Queensland Plan (QueenslandGovernment, 2018).

Similarly to Queensland and ACT, the Victorian state government has set ambitious targets for the integration of RE. The state made a commitment in June 2016 with a target of 25 percent of RE generation to be sourced from RE, to be achieved by 2020 (VRET), and 40 percent by 2025. The VRET is supported by the Victorian Renewable Energy Auction Scheme (VREAS), which provide the 'support agreements' to successful bidders for a term of 15 years. The mechanism by which the scheme will be promoted is a reverse auction, utilising a contract for difference approach to be incorporated as part of the support agreement. The proponents will compete against each other in their bidding for the base tariff and base amount to be incorporated in the contract for difference. The state is then responsible for any shortfalls in electricity pool price being lower than the base tariff, and any excess pool price differences being paid back to the state. The 2017 VRET reverse auction kicked off with request for proposals for the procurement of 550 MW of technology neutral RE and up to 100 MW of large scale solar, totalling 650 MW in the aggregate (VicGov, 2018a).

2.5.5   Transition to Renewable Energy

Australia's electricity market functions as an energy only market. I.e. there are no capacity payments to generators for providing reliability of supply. The market functions as a continuous auction of electricity in a wholesale pool. In order for the market to be sustainable, it must promote the optimal mix of generators which can cover their fixed capital costs and variable operational costs (Nelson, 2018). Blazquez *et al.* (2018) points out that the full decarbonisation of the market, with 100 percent penetration of RE would result in a market collapse given the almost zero marginal cost of RE. Conventional dispatchable technologies provide price signals which the wholesale price is based on. Removing this would ultimately lead to a shift in market philosophy from free market to administered subsidy (Blazquez *et al.*, 2018). The shift in a market from low cost coal fired power to zero marginal cost RE generation, which will ultimately displace conventional generation in NEM dispatch has been reiterated in AEMO's Integrated Systems Plan released in 2018 (AEMO, 2018a). Potential solutions to this dilemma are to switch to a centralised non-competitive market or to subsidize conventional generators in the form of capacity payments. Both of these options mark a move away from a free-market based system (Blazquez *et al.*, 2018). Nelson (2015) goes on to make the point that the current market is already constrained in reality and is not a theoretical free market, with the existence of market price caps, financial intervention, and regulatory interference. Nelson (2015) also concurs with Blazquez *et al.* (2018) that a substantial increase in penetration of RE would theoretically result in the collapse of the market structure, and would require interventions in the form of increased subsidy for the continued development of RE (Nelson *et al.*, 2015).

There is approximately 75 percent of the thermal capacity in the form of coal and gas in Australia that is passed its original design life (Nelson *et al.*, 2017). These plants will have to be replaced in the near future, most likely with RE to ensure that the energy capacity continues to meet demand in a sustainable way. The retirement of coal fired power stations has been shown to be disorderly, with a lack of planning for new capacity to replace the retired capacity. This is reflected in dramatic wholesale price increases upon closure of such plants (Nelson *et al.*, 2017). It is argued by Nelson *et al.* (2017) that a regulatory closure policy which would be based on generator age, could potentially address political concerns around pricing volatility. Upon the planned withdrawal of firm capacity, the suitable replacement generation would need to be incorporated into the market. The conventional cost recovery mechanisms of coal fired power plants, which recover costs during peak priced revenue periods of high demand is potentially at risk due to the low marginal costs of increasing RE that may be operating at peak times. Price volatility would therefore have to become extreme in markets to cover fixed costs of conventional power plants (Nelson *et al.*, 2017). Blazquez *et al.* (2018) agrees, stating that at the point that market pool prices fall below the marginal operating costs of conventional plant, that this would mark the failure of the free market, and that capacity payments, investment support and mandates would potentially be required to avoid conventional plants being mothballed. This is referred to as the 'blend wall' at which point market pool prices drop to a point which is below the marginal cost for conventional plant (Blazquez *et al.*, 2018).

The scenario in Australia has been compared to other markets such as the U.S., where there has been similar closures of aging coal fired power stations. The U.S. has managed to transition through this period without the volatility in pricing as witnessed in Australia. The major differences however between the two markets include a stable policy climate in the U.S. and the cost of natural gas. There has been a lack of investment in gas fired generation in Australia due to policy instability and high costs of gas due to the current export market. Dispatchable plant is required to compliment the increased levels of RE generation. Gas is preferred to coal due to its reactionary times and flexible capabilities. The generation mix will be dependent on the price of gas in the future (Nelson, 2018).

Nelson *et al.* (2015) argues that policy should guide investment to a generation mix that makes financial sense. The current policy goes against this sentiment. The disorderly retirement of aging coal fired power plants is resulting in rising energy prices. Energy policy needs to address the mechanisms required to promote an optimal energy mix, and how the RET should be structured. The current structure of the NEM being an energy only market requires review (Nelson *et al.*, 2015).

2.5.6   Integrated System Plan and Grid Access

The Australian Electricity Market Operator (AEMO) released the Integrated System Plan (ISP) in July 2018 (AEMO, 2018a), outlining the proposed future energy mix and requirements in terms of grid strengthening and transmission expansion that would be required. The ISP sets out a framework for what is required in terms of energy mix and energy security and reliability. Achieving these goals requires the appropriate policy to be implemented. The ISP foresees a shift away from coal fired generation to a market dominated by RE (wind and solar PV), with support from gas fired generation and energy storage such as pumped hydro. This is already observed by the record levels of RE that have been committed, with 5 GW of RE generation expected to be operational in the next two years. Coal fired power stations are expected to be increasingly operated at lower levels due to the merit order effect of RE. Gas generation is expected to reduce initially but increase to previous levels as coal generation exits the market (AEMO, 2018a).

The ISP acknowledges a dramatic shift in the energy market, with the introduction of a record capacity of RE, that is installed and committed to be installed over the next few years. It is noted that although there has been an increase in economic and population growth, this has not translated into an increase in electricity demand. This is partly due to the increasing quantity of rooftop solar and energy efficient devices (AEMO, 2018a). With increasing network costs contributing to approximately half of the Australian customers electricity bill, an increasing number of Australians are switching to rooftop solar to avoid such costs. This is explained as a death spiral, as more customers leave the network, the remaining customers are burdened with increasing costs, which prompt more customers to leave. There is evidence in the NEM that such a scenario has unfolded (Nelson *et al.*, 2017). The ISP outlines a significant increase in network infrastructure to accommodate RE, which may exacerbate the issue. While the demand is reducing, supply has also seen a reduction with aging coal fired power plants requiring replacement with new capacity to meet consumer demand (AEMO, 2018a).

The ISP states that the intention is to maintain existing coal fired power plants up to the end of their design life. Upon the retirement of these plant, the most cost effective solution is the replacement with generation capacity which consists of a portfolio of RE (wind and solar), energy storage such as pumped hydro, flexible thermal generation (i.e. gas generation) and expansion of transmission networks (AEMO, 2018a). It is noted that there is approximately 75 percent of the thermal capacity in the form of coal and gas in Australia that is passed its original design life (Nelson *et al.*, 2017), which contradicts the statement in the ISP. The ISP notes the need for an increasing amount of system storage in the medium term to provide additional system security and reliability. It is proposed that this be included as part of new RE generating plant (AEMO, 2018a). The need for developing cost effective solutions for energy storage is an important area of focus for an energy system that will be characterised by high levels of RE penetration (Hua *et al.*, 2016). This forecast for the future energy mix reflects what is discussed in the literature in terms of replacement of aging coal plants with RE and flexible gas fired power plants (Nelson, 2018). The implementation of the ISP through appropriate energy policy may bring some order to what Nelson (2018) describes as a disorderly retirement of aging coal fired power stations. As Nelson *et al.* (2015) points out, the development of energy policy and a rethinking of the NEM would also be required alongside the transition to RE (Nelson *et al.*, 2015), which is outlined under the ISP.

The increase in RE generation is expected to result in more dispersed generation, which will require transmission networks to extend to locations currently not serviced. It is acknowledged that a much larger grid infrastructure will be required to take advantage of locations with superior RE resources. Additional interconnectors are proposed to establish greater transfer capability between Victoria and South Australia, NSW and South Australia, and Queensland and NSW (AEMO, 2018a). This is reiterated by Hua *et al.* (2016), Australia does not have sufficient network capability and it is imperative that network upgrades are carried out as soon as possible so that the RE development is not stalled in future. The barrier to grid infrastructure expansion is noted to be cost related as opposed to technical constraints (Hua *et al.*, 2016). This is discussed by Nelson's description of the 'death spiral' with regards to exorbitant network costs as explained above (Nelson *et al.*, 2017).

With the retirement of thermal generation and increasing RE generation, additional services will be required. The forecast for 2040 is for a reduction of 16 GW of synchronous (conventional) generation, and an installation of 38 GW of RE. This is expected to have a significant effect on system strength. Technical solutions are required to address the loss of these services, with the significant increase in non-synchronous generation. It is identified that support in the form of technical solutions to provide voltage control, frequency management, system strength, dispatchability and power system inertia will be required. It is possible for these services to be procured from the market as well as being provided as a requirement for RE plant connecting to the grid. In South Australia specifically, the ISP has identified that synchronous condensers will be required to supply system strength and inertia. Furthermore, system restart ancillary services (SRAS) which are currently provided by coal fired generation will be required as RE increases its market share. RE plants do not inherently provide such services, although pumped hydro has shown to have grid forming capabilities and could potentially be used for this purpose. SRAS may also be potentially provided by RE plants with the provision of specific equipment such as grid forming inverters. This provides a non-synchronous plant with synchronous capability (AEMO, 2018a). With the increasing volume of RE into the NEM, and retirement of conventional generation, an increasing number of mitigating factors will be required to ensure system operations are not compromised.

Grid connection costs can be a significant barrier to projects being realised, and are not standardised across projects. They can range from shallow connection to deep connection charges, depending on the point of connection. This can be especially limiting for medium sized projects, which do not have the economies of scale (Byrnes *et al.*, 2013). Connection costs are expected to increase and become an even more significant factor as the market shifts to greater penetration of RE, resulting in an increased requirement for grid related technology (Byrnes *et al.*, 2013). The ISP is proposing the development of Renewable Energy Zones (REZ), which aim to distribute the costs of the grid strengthening solutions between the various energy generators that are connecting to the gird in that region. From 2030 onwards, it is expected that most RE developments will require some sort of system strength remediation in order to connect to the grid. It is expected that contributing towards providing system strength for a REZ will be more economic than providing system strength for each individual RE project (AEMO, 2018a). The question whether network costs should be socialised to a degree instead of being borne entirely by the RE generators is something that may have to be considered by government (Byrnes *et al.*, 2013).

The action plan outlined under the ISP proposes immediate expansion and interconnection of the transmission infrastructure, reduction in the congestion for RE projects in western and north west Victoria, and finally to remedy the issues with system strength in South Australia. The second step would be to then invest in the infrastructure required to provide the further proposed interconnections between the eastern states (AEMO, 2018a). It is recommended that coordination is improved between national government and the state governments in order to achieve the required system upgrades and interconnections between states, that is required for the integration of further RE penetration in Australia (Hua *et al.*, 2016).

2.6   CHAPTER SUMMARY

This chapter reviewed the literature on global approaches to clean energy policy, detailing the specific regulatory and fiscal policy methods employed to promote RE development. The merits and significance of the various energy policy schemes were investigated. Specific barriers to RE development were identified such as grid access, political transparency and technological challenges.

The literature was then investigated in the context of Australia. The status quo of RE development in the context of Australia's energy policy was reviewed. The potential barriers and requirements for a transition to an energy mix with a greater proportion of RE was then investigated.

3.1   INTRODUCTION

This section outlines the approach to carrying out the research. The methodology is outlined in Figure 3-1. The research question and literature review influenced the selection of research method employed. A qualitative research approach was selected using a case study method. A single case study was selected. The case study investigated was the renewable energy division of a leading global consultancy operating in Australia. The unit of analysis were defined as the senior management staff employed in the renewable sector at such consultancy. The approach for data collection is detailed, with interviews being the preferred method of data collection. Standardised open ended interviews were considered most appropriate. The interview questions and theoretical patterns were identified based on the review of the literature. The individuals were interviewed and the data transcribed. The data was analysed by employing a method of pattern matching. The observed data from the interview process was then analysed and compared against the expected pattern(s).

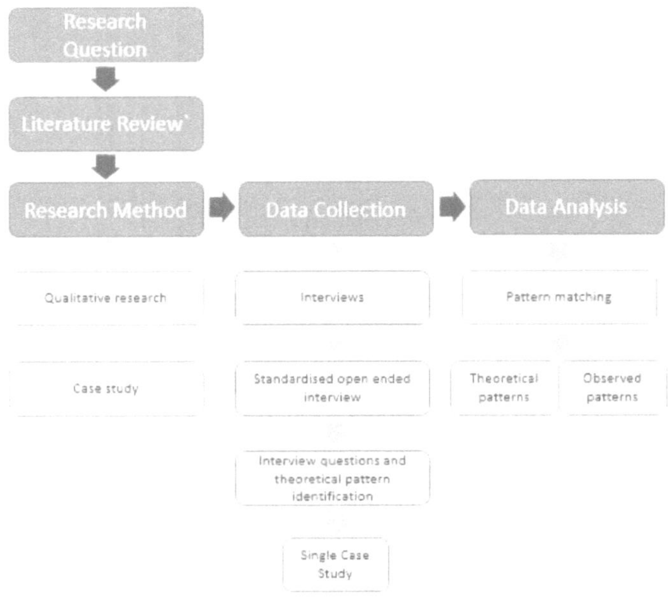

Figure 3-1: Research methodology

There exists two general approaches to research, which is qualitative and quantitative research (Kothari, 2004). Qualitative research is based on qualitative phenomenon, as opposed to quantitative research which is based on measurement applicable to phenomenon which can be measured in quantity (Kothari, 2004). Qualitative research involves the subjective analysis of the data that emerges from the research, such as opinions and behaviour (Kothari, 2004). Investigation of the research question involves understanding political context, technical issues and market behaviour which are considered to be complex as they are dynamic and result in emergent phenomenon. A qualitative approach is therefore the preferred method of analysis to be employed in addressing the research question.

Qualitative research is associated with research methods such as unstructured interviews, focus groups and participant observation. Traditional approaches in undertaking qualitative research include ethnographies, case study research, and grounded theory (Lund, 2012).

Case study research is a method that can be used to compare observed data to general theory to contribute to knowledge. It involves the detailed investigation into a phenomenon and understanding of its context (Cavaye, 1996). A case study approach has been suggested, according to Yin (2003) as being applicable when the following phenomena exist; The study seeks to answers 'how' and 'why' questions; the behaviour of those involved in the study cannot be manipulated; The greater context is required to understand the questions being posed; clear boundaries do not exist between question and the context in which it is being asked (Yin, 2003).

The research question is to identify the barriers and support mechanisms which may impact on the growth of RE in Australia. To address the question, it was important to understand the greater context of the National Electricity Market and energy policy in Australia, which interplays with the development of RE. Yin (1994) classifies case studies as being an applicable method when studying current phenomenon of a real-life nature. This is applicable to the development of RE in Australia, and the case study method was therefore the selected approach employed to carry out the research.

A qualitative case study is an approach that allows for the analysis of multiple sources of data. This allows the researcher to analyse the data through multiple lenses, and formulate a broader understanding of the various phenomena that may exist (Baxter and Jack, 2008). There are a number of approaches that can be employed when carrying out case study research, and there is no single accepted way of conducting case study research (Cavaye, 1996). The type of case study which will be carried out must be determined. According to Yin (2003) and Stake (1995) there are a various types of case studies which can be conducted. Yin (2003) identifies various case study types, including explanatory, exploratory, descriptive and multiple-case studies. Stake (1995) classifies case studies as being collective, intrinsic or instrumental. Table 3-1 details the various case study types and descriptions for each type.

Table 3-1: Case study types, adapted from (Baxter and Jack, 2008)

Case Study Type	Description
Explanatory	An explanatory case study seeks to address a question that is attempting to identify causal links as a result of real-life interventions. These causal links would be too complex to be understood by survey or experimental strategy (Yin, 2003).
Exploratory	Exploratory case studies are the preferred method when the interventions being assessed have no definitive effect or outcome (Yin, 2003).
Descriptive	A descriptive case study is preferred when an intervention or phenomenon is explained in a real-life context (Yin, 2003).
Multiple-case studies	The multiple case study approach seeks to identify contrasting views or similarities between cases. Comparisons are drawn between cases, highlighting similarities or areas of divergence (Yin, 2003).

Intrinsic	An intrinsic case study is a term used by Stake (1995) which describes a scenario in which the case itself is of interest and the purpose is not to infer information from the case in order to build some general theory or gain insight into a generic phenomenon (Stake, 1995).
Instrumental	Contrary to an intrinsic case study, an instrumental case study is an approach selected to provide insight into a situation, and the findings applied to confirm or refute a theory. The case itself is not of significant importance, but rather provides information to the researcher to develop a viewpoint for an external interest (Stake, 1995).
Collective	Collective case studies are comparable to multiple case studies, as described by Yin (2003).

Yin (1994) describes a case as being composed of information (unit/s of analysis), which could comprise an individual, document, artefact etc. The case represents the topic or setting that is being investigated, such as an organisation. The investigation of a single case can be designed so that multiple units of analysis are embedded within the case (Cavaye, 1996). A case is usually a thing or an entity that can be visualised, and can range from a person to a document (Stake, 1995). The case which represents the topic or item under study, such as an organisation may contain more than one unit of analysis. By comparing the data from various units of analysis within a single case, a theory can be developed. Single case studies with more than one unit of analysis is comparable to a multiple case study approach (Cavaye, 1996). In the same way that a multiple case study approach seeks to identify contrasting views or similarities between cases, highlighting similarities or areas of divergence, a single case study with more than one unit of analysis can be employed to the same effect. Yin (1994) refers to this as an embedded design. Cavaye (1996) concludes that the embedded design of a single case study is comparable to a multiple case study. This is confirmed by Baxter and Jack (2008), who note that a single case with embedded units is similar to a multiple case study. Although it is noted that the context is different between cases, whereas the embedded units will be analysed under the same setting (Baxter and Jack, 2008).

3.3.1 Case Study Design and Unit of Analysis

As summarised by Yin (2003), the multiple case study approach seeks to identify contrasting views or similarities between cases. Comparisons are drawn between cases, highlighting similarities or areas of divergence. A multiple case study (as opposed to a single case) is argued to provide more robust findings which have been validated by converging sources from multiple cases (Baxter and Jack, 2008). By studying a single case, an in depth description can be obtained which provides a basis for theoretical constructs to be developed (Cavaye, 1996). Yin (1994) recommends a single case approach when the case is considered to be able to provide critical evaluation of existing theory, or for unique events. Cavaye (1996) also notes how a single case study can be employed to confirm or refute theory.

The research question aims to identify potential barriers and support mechanisms which may impact on the growth of RE in Australia, from the perspective of energy specialists at a leading consultancy operating in the built environment in Australia. The energy division of leading consultancy in the built environment was identified as being unique and suitable for the purposes of being used as a single case study, to address the research question.

The organisation selected is one of the largest tier one consultancies in the built environment in Australia, and globally, with global employees totalling 37,000. The organisation's renewable energy division has global presence, and operates in the renewable markets most notably in Canada, USA, England and Australia and New Zealand. The organisation has significant presence across the power and energy markets in Australia, and is involved in the development, construction and operational phases of the RE lifecycle. The organisation is not only focussed on RE, but covers the entire power sector, including other forms of generation such as coal and gas, as well as providing services to network services providers who own distribution and transmission assets, and undertaking power system modelling, which requires in depth knowledge around the National Electricity Rules and the NEM. In

Australia, the power sector consists of 90 employees. These personnel are distributed throughout Australia, with the majority located in Melbourne, Sydney and Brisbane offices. The type of projects which the RE division focus on include; pre-feasibility assessments of all generation types, including providing services specifically for ARENA; wind and solar resource and energy yield assessments, pre-financial close and acquisition technical due diligence on behalf of investors and lenders, including the CEFC. This work covers a broad scope from generation technology, civil and electrical design, contractual arrangements, financial model assumptions, grid connection assessment and grid system constraints, asset management arrangements, power purchase agreements, energy yield assessment, and environmental approvals review; the organisation provides development support services to developers, from environmental and planning services relating to project development approvals, to contract specification development and tender support; during construction some of the main services include engineering services to lenders, developers and contractors depending on the scope required; during operations services provided include asset performance verification testing and operational plant performance and energy assessments.

The organisation is currently providing ongoing services to numerous clients representing multiple gigawatts of combined energy capacity, specifically in the RE sector, which is supported by a renewables team of nearly 30 engineers.

As discussed the consultancy is involved in all phases of the RE lifecycle and has interactions with all stakeholders associated with these phases, including project developers, network service providers, contractors, AEMO, generation offtakers, landowners, state planning authorities, lenders, investment managers and community. There are other large global consultancies in the built environment, which are active in the RE sector in Australia, although it is debatable whether any other organisation can provide the full suite of services as outlined above. The organisation, and particularly the energy division is therefore considered to be a unique case study in the context of the relevant RE knowledge base that is embedded within the division.

Senior management personnel were chosen as the unit of analysis, the justification of this selection is further discussed under Section 3.4. The participants have been purposefully selected based on their experience and knowledge to address the research question of what the potential barriers that may need to be addressed and strategies or policies that may be required for continued growth of RE in Australia. They are also considered to be specialists in the energy sector in Australia, which also addresses the research question which is limited to the perspectives of such energy specialists. They are considered to be energy specialists because of their links across the business in Australia as well as their industry connections that they are required to maintain across the business globally. The depth of their overall knowledge of the industry in Australia in their specific field as well as well as across the market generally is required to be extensive, as dictated by their position. The chosen personnel have been involved in the industry in Australia for a significant amount of time, and understand previous challenges faced by the industry as well as potential future constraints.

As such, a single case study, being a leading global consultancy, active in the Australia RE industry was considered to be an appropriate approach. The intention of the research has been to evaluate the existing theory by collecting data from multiple respondents, being the senior level employees (unit of analysis) working in the renewable energy sector for such consultancy. This approach has allowed for critical evaluation of existing theory by way of specialist individuals providing detailed information pertaining to the RE industry in Australia.

3.3.2   Case Study Considerations

Case studies are subject to the individual researcher's subjectivity in terms of their own influence during the data collection and analysis process. The analysis relies on the researchers interpretation of the information gathered during the case study, which may imply limitations on research findings (Darke *et al.*, 1998). This limitation is explored by Flyvbjerg (2006) who refers to one of the misunderstandings of case study research, which is that the method is subject to the researchers bias and that findings are interpreted and validated, to confirm the preconceived notions of the researcher. It is argued the case study method is no more influenced by the researcher's subjectivity than other research methods may be influenced. On the contrary, case study research may contain greater bias towards falsification of a researcher's preconceived notions, rather than verification (Flyvbjerg, 2006).

A limitation of case study research is that it is not possible to isolate independent variables which may restrict the internal validity of the study's conclusion (Cavaye, 1996). Although this is a drawback, the credibility of data is improved by the use of multiple sources of data. The data sources are integrated and converged during the analysis to create a more informed understanding of the phenomenon under study, providing greater credibility (Baxter and Jack, 2008).

A common criticism of case study research is that its dependence on a single case can result in limitations on generalising the conclusion to the population (Tellis, 1997). Yin (1994) argues that whether it is a single case, two cases or many more cases, this does not transform the case study into a macroscopic study, therefore even a single case is acceptable (Tellis, 1997).

3.4   DATA COLLECTION

Case study research can be conducted by obtaining evidence for analysis. Such evidence can comprise documents, archival records, direct observations, participant observations, physical artefacts, and interviews (Yin, 1994). Case studies often involve the use of multiple sources of data, which can result in the collection of vast quantities of data which then becomes a complex task to manage and analyse (Baxter and Jack, 2008). Not all sources will be relevant for all case studies, and it is the researcher who will have to select which data sources to investigate (Yin, 1994).

Documents could refer to letters, articles or any other document that may be relevant to the investigation. Documents can provide evidence of events, and can be used as a source of triangulation to validate data from other sources (Tellis, 1997). Documents may have been relevant to this particular investigation. The most relevant may be news articles or government publications. It is noted that these may be subject to bias, based on political objectives and may lead the researcher astray. The validity of such sources may be questionable in the context of energy policy and the debate surrounding the matter. Documents have therefore been excluded as a source of data. Archival records, which can be organisational records, survey data etc. (Tellis, 1997) were not considered to be relevant to the study. Direct and participant observation involves the researcher making observations of behaviours or collecting data, and either being excluded from the group being observed (direct) or forming a participatory role (participant) (Tellis, 1997). The timeframe for observing and collecting data which would be applicable to the research question would be extensive. This was not considered to be appropriate in addressing the research question at hand. Physical artefacts as the name implies refers to items collected such as tools or objects which are relevant to the study (Tellis, 1997), this too was not considered relevant for the topic under investigation.

Interviews are considered to be one of the most important techniques for the collection of evidence in a case study investigation (Tellis, 1997). Interviews were considered to be relevant to the investigation, with the ability to illicit insight from the sources under investigation. They provide in-depth information, based on participants experience and knowledge surrounding the area under investigation (Turner III, 2010). Interviews allow for the discussion around key areas identified in the literature, and the subsequent insight into such knowledge areas. Other sources of data were not considered to be as useful or relevant to the investigation, and interviews comprise the sole source of data gathering.

Participant selection has been based on a purposive sampling technique. It is a non-random technique which does not require a set number of participants, and is a conscious determination of which participants to incorporate into the research based on the relevant knowledge and experience that the participants possess (Etikan *et al.*, 2016). The number of participants should be relevant to the addressing the research question. A small number of participants who are purposefully selected, based on their exposure to and experience of the topic under study, could produce highly relevant data for analysis (Cleary *et al.*, 2014). Purposive sampling involves the selection of participants on the basis that they can contribute to addressing the objectives of the study (Collingridge and Gantt, 2008). The research question focusses on the perspectives of energy specialists at a leading consultancy operating in the built environment in Australia. A conscious selection of participants who are considered to be specialists in the topic under study has been made based on their position within the company, and the relevant knowledge and experience they possess. The companies power division in Australia and New Zealand is led by the Director of Power, who oversees all of the divisions under Power, across the region, including the renewable energy division. The Power division includes three Section Managers who oversee each division in the Power sector, who report directly to the Director of Power. The Section Manager for the renewable energy division oversees that division entirely. It is noted that Section Manager is also a member of the ARENA panel, and has valuable insight into the RE market and future RE developments in Australia. The renewable energy business unit consists of a wind engineering and solar engineering division. Each unit is overseen by the two Engineering Managers, who report directly to the Section Manager. There is also a focus on new energy, such as energy storage and hydrogen which is led by the Engineering Manager for New Energy. Engineering disciplines such as civil, electrical and grid connection are led by the relevant Engineering Managers for each discipline. The most relevant participants, considered to be specialists, based on their position within the company that could contribute to the topic under study were therefore considered to be the Director of Power, the Section Manager and the Engineering Managers for Wind, Solar, New Energy and Electrical. Interviews have been carried out with these participants who have been purposefully selected from the larger pool of employees (100) available within the Power sector in Australia. There are various interviewing methods that can be employed, which is discussed in the next section.

3.4.1   Interview Techniques

The literature covers a variety of interview techniques, Jennings (2005) makes a comparison of structured, semi-structured and unstructured interviews in terms of style, design, researcher stance, exchange and data collected. Similar interview techniques may be labelled differently, as demonstrated by Jennings (2005) who lists the types of unstructured interviews, providing descriptions and synonyms where applicable for each type.

Turner III (2010) describes three different approaches when carrying out interviews for data collection. These are referred to as;

- Informal conversational interview

- General interview guide approach

- Standardised open ended interview

The informal conversational interview is characteristic of an unstructured interview process, in which the interviewer has not developed any questions prior to the interview. The setting may be one in which the interviewer is immersed in an experience and the questions emerge as part of the experience as a response to what is observed (Turner III, 2010). This interview approach can be compared to an unstructured interview or ethnographic interview (Patton, 2002). This approach is considered to be more suitable to a situation in which the subject under investigation forms part of the interview process, which illicit stimuli to spark conversation, for instance a cultural ceremony. The technique was therefore not considered to be appropriate.

Turner III (2010) describes the general interview guide approach, which is a more structured interview process than the previously discussed informal conversational interview. A set of questions is developed as more of a guideline, in which the interviewer can interpret such questions and pose to the interviewee in a manner they see fit. This allows the interviewer to modify the questions during the interview, based on the responses received. The interviewer can also follow up with subsequent questions not originally formulated. Similar interview techniques, based on the description provided by Jennings (2005) may be the focussed interview, non-schedule standardised interview, unstructured schedule interview, and semi-structured interview. It is a conversational approach employing a structured and semi-structured format. It is applicable to validate information and elicit experience and opinions from the participants (Jennings, 2005). The advantage of such technique is that flexibility is allowed to foster an interactive discussion, although greater structure is provided compared to an informal conversational interview. The disadvantage is that depending on how the interviewer poses the question, a different response might be elicited from the participant. This may also introduce greater researcher bias during the process (Turner III, 2010). Data analysis may also be challenging, given the potentially broad spectrum of discussion from different participants. This method could be employed, as it will allow for a guideline of standardised questions to be posed, with the ability of the interviewer to elicit further information from the participant if required, or to delve deeper into the topic of discussion based on the participant's response. However, this was not selected as the preferred method, as the literature has provided specific areas of focus to verify and there is less of a need to modify questions from participant to participant.

A standardised open-ended interview provides the greatest structure and is one of the most popular interviewing methods employed. This is due to its structure, in that identical questions are asked which removes the disadvantage of varying responses of the general interview guide approach, but the questions are posed so that the answers provided are open ended. This allows the participant to provide detail that may have otherwise been missed, and also reduces bias that the interviewer may impose on the interview. The open ended interview allows for the participant to provide as much detail as they wish, and it allows for the interviewer to follow up with more detailed questions based on the response (Turner III, 2010). The technique shares similarities with the formal interview, defined as an interaction which has a fixed agenda, characteristic of a set of structured questions which may include questions of an open-ended nature (Jennings, 2005). A possible disadvantage may be that the open ended responses may be difficult to analyse, given the potentially varying levels of detail and responses from each participant (Turner III, 2010). The literature surrounding the research question allowed for specific questions to be formulated which could be posed to participants to obtain insight into each topic. This method was considered to be the most appropriate, as it allowed for standardised questions to be posed, as well as follow up questions based on the participants' response, without any restriction on participants.

3.4.2   Interview Questions

Interview questions have been formulated, which relate to the potential barriers and questions around policy support mechanisms which may have an impact on RE growth. A set of seven questions are listed. Each question represents a topic which has been identified in the literature as being relevant to addressing the research question. The questions have been kept brief in order to elicit but not influence the participants' response. The relevancy of the interview question is justified by the sources listed under Table 3-2, which have been referenced in the literature review and discuss the topic in detail. The purpose of each question is to elicit a response from the participant to empirically test the theoretical patterns which have emerged from the literature review. The method of data analysis is discussed in more detail under Section 3.5.1. The interview questions, as well as the justification for the question, literary sources and identified pattern is detailed in Table 3-2.

Table 3-2: Interview questions

	Question	Justification	Sources	Pattern
1	The national policy is that of a Renewable Portfolio Standard with LGCs, compared to a FIT system which is being employed at state level with contract for difference. Should this be the preferred policy?	To ascertain the effectiveness of the chosen policy and determine whether other energy policy may be more suited to RE development	(Simshauser and Tiernan, 2018) (Byrnes and Brown, 2015) (Diaz-Rainey and Sise, 2018) (Aparicio *et al.*, 2012) (Buckman *et al.*, 2014) (QueenslandGovernment, 2018)	Energy policy – preference to FIT to promote RE at the utility scale.

			(VicGov, 2018b)	
2	How important will national energy policy be for continued RE integration, assuming continued state support for RE integration	To determine the role and importance of national energy policy for continued RE development	(Simshauser and Tiernan, 2018) (Byrnes and Brown, 2015) (Diaz-Rainey and Sise, 2018) (Aparicio *et al.*, 2012) (Buckman *et al.*, 2014) (QueenslandGovernment, 2018) (VicGov, 2018b)	Energy policy – Long term national policy is critical for RE growth.
3	Is successful RE development dependent on financing institutions such as the CEFC and ARENA, based on the number of projects that these bodies finance?	To assess the importance of fiscal initiatives such as government loans and grants to continue the development of RE	(Diaz-Rainey and Sise, 2018) (Byrnes *et al.*, 2013) (Byrnes and Brown, 2015) (Hua *et al.*, 2016)	Energy policy – Fiscal support in terms of national funding agencies is effective in promoting growth of RE.
4	Connection costs can be a barrier to entry, especially small to mid-sized projects.	To understand how RE projects will overcome grid connection costs as	(AEMO, 2018a) (Nelson *et al.*, 2017)	Technical – Increased grid connections costs

	Assuming increased penetration of RE and increasing grid costs required to meet performance standards (synchronous condensers and reactive plant), how might RE projects overcome this?	a current barrier to entry.	(Hua *et al.*, 2016) (Byrnes *et al.*, 2013)	expected to become a barrier to entry.
5	Renewable energy zones (REZ) have been highlighted in the ISP as a potential solution to reduce costs associated with network infrastructure. Is this likely to be conducive to RE development?	To determine the effectiveness of the development of REZs	(AEMO, 2018a)	Technical – REZ highlighted as a potential solution to increasing grid constraints.
6	The ISP refers to a future energy mix dominated by RE, storage and gas powered generation. How likely is this scenario given high gas prices?	To understand the likelihood of the scenario put forward under the ISP.	(AEMO, 2018a) (Nelson, 2018)	Electricity market – Future proposed generation mix considers gas fired generation to supplement a

				RE mix, but may be constrained by gas pricing.
7	How suitable is the NEM to deal with the future of a significant proportion of RE, given the zero-marginal cost of RE and its effect on wholesale pricing? Is market reform necessary?	To ascertain whether the current NEM will be a suitable platform for the shift towards significant incorporation of RE.	(Blazquez *et al.*, 2018) (Cludius *et al.*, 2014) (Byrnes *et al.*, 2013) (Diaz-Rainey and Sise, 2018) (Nelson *et al.*, 2017)	Electricity market – A shift towards greater allocation of RE will require a restructure of the NEM.

3.4.3 Participant Labelling

Anonymity in research refers to the process of ensuring that research participants identity is not disclosed. The anonymity of the research participants should be considered when undertaking research, such as interviews (Clark, 2006). The anonymity of the participants has been ensured by coding each participant as referred to in Table 3-3.

Each participant is labelled according to their level and type of education and number of years of experience in the energy sector. Participants with a commerce and engineering educational background are classified as *Comm.* and *Eng.* respectively. An example of the coding is shown in Figure 3-2.

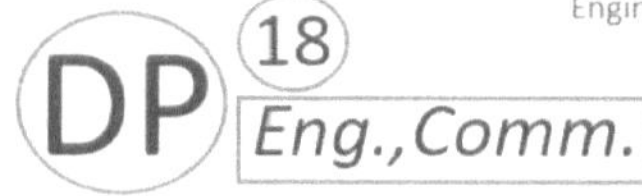

Figure 3-2: Example of participant labelling

Table 3-3: Participant labelling

Participant reference	Description
DP[18]Eng., Comm.	Director: Power
SMR[17]Eng.	Section Manager: Renewables
EMW[14]Eng.	Engineer Manager: Wind
EMS[12]Eng.	Engineering Manager: Solar

3.5 DATA ANALYSIS METHOD

There are various approaches to data analysis. Miles and Huberman (1994) promote a use of matrices and networks to assist in the interpretation of qualitative data. Their argument is that pages of text from interview transcripts can be tiresome and provides data sequentially instead of simultaneously comparing multiple variables. Yin (1994) suggests a few analytical techniques that can be employed as part of case study analysis, such as pattern matching, explanation building and time series analysis; Pattern matching aims to compare empirically founded patterns with predicted patterns (Yin, 1994). Pattern matching requires the researcher to observe and attempt to identify a link between a theoretical/expected pattern and the emergent pattern from the data (Trochim, 1989). This technique was considered to be most appropriate for the analysis of the research data compiled from the various interviews. Using a pattern matching technique allowed findings from the literature to be compared to the patterns that emerged from the data; Explanation building is an explanatory technique which involves multiple iterations of revising a theoretical statement based on findings of multiple case studies (Yin, 1994); Time series analysis requires the analysis of a sequence of events which are expected to lead to a theoretical outcome, and involves asking certain 'how' and 'why' questions (Yin, 1994). Explanation building and time series analysis were not considered appropriate techniques in the context of the research question.

3.5.1   Pattern Matching

Campbell (1975) describes the pattern matching approach which can be used as part of a single case study design in which known theoretical propositions are compared with the observed characteristics of the case study. Pattern matching has been selected as the preferred technique to compare the theoretical data (theoretical patterns which emerged as part of the literature review) which forms the basis of the interview questions, to the observed data (observed patterns which emerged from the interview process). A pattern is an arrangement of objects or entities, which are not random in nature and can be described (Trochim, 1989). Pattern matching is the process of linking an observed pattern and a theoretical pattern. This is illustrated in Figure 3-3, which describes the process of pattern matching. Steps involved include, developing the theoretical patterns based on theory and in this case, based on ideas presented in the literature. This is depicted in the top half of the illustration. The next step is gathering data, by conducting interviews and recording transcripts in this case, which is then organised and analysed to identify patterns that emerge. Finally the observed pattern from the data is compared with the theoretical pattern to ascertain whether there is a match (Trochim, 1989).

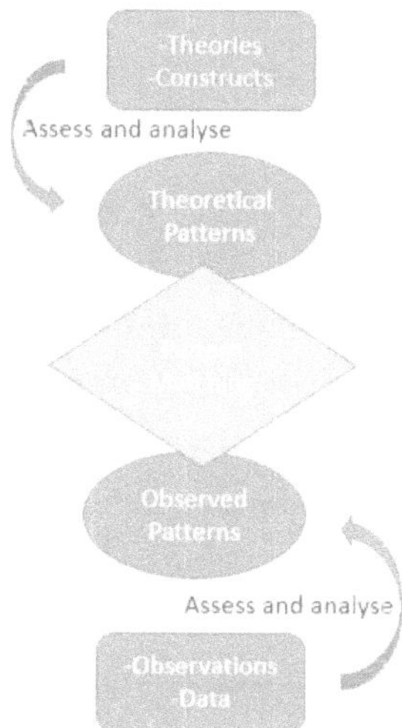

Figure 3-3: Pattern matching process. Adapted from (Trochim, 1989)

Pattern matching is similar to testing a proposition, in that the development of a theoretical pattern is effectively a proposition of what could be expected to be observed. The data is analysed to assess the validity of the theoretical pattern. The key difference is that pattern matching encourages the development of more complex propositions, which may contain multiple variables (Trochim, 1989). Should the patterns match, the internal validity of the study is strengthened (Tellis, 1997). Should the expected pattern and the observed pattern not match, this may point to the theory being incorrect or the data gathered being inaccurate. Pattern matching is a deductive procedure that involves the comparison of the observed case data (interview transcripts) against the predictions of the theory identified in the literature (Hyde Kenneth, 2000). The process of pattern matching involves the comparison by way of judgement of the theoretical patterns against the data captured in interview transcripts to ascertain whether there is a match (Hyde Kenneth, 2000).

A pattern match strengthens the theory, so long as there are no other theoretical patterns which may result in the same outcome (Trochim, 1989). It is also important to ensure the theoretical validity of the study is demonstrated. In terms of pattern matching, this can be done by employing a negative case sampling strategy. This refers to searching the data for patterns which emerge that do not fit with the expected outcomes or theoretical patterns (Johnson, 1997).

3.6   CHAPTER SUMMARY

This chapter has outlined the preferred research methodology, which is a qualitative research methodology. The method deemed to be most suitable for conducting the research was a case study approach, in which a single case study was analysed with an embedded design. Senior level employees were selected as the units of analysis working in the renewables sector at a leading engineering consultancy in Australia (the case). The preferred data collection method was motivated to be face to face interviews. A standardised open ended interview was the preferred interview type that was selected. The data analysis methods were explored, and pattern matching was selected as the preferred technique. Theoretical patterns were identified in the literature, and interview questions formulated, to elicit data with regards to each of the identified patterns.

This Chapter presents the findings of the six interviews conducted as part of the data collection process. The data gathered from the responses to each of the interview questions has been presented under the relevant theoretical pattern outlined in Table 3-2, as part of the pattern matching technique discussed in Chapter 3.

The patterns which have emerged from the literature review and are referenced in Table 3-2 include;

- Energy policy: The vital importance of a stable national energy policy which includes a selection of regulatory instruments such as FITs, RPS, and fiscal initiatives as a supplementary policy.

- Technical: RE development will depend on addressing technical issues associated with grid connection and stability, which will have an associated cost to overcome.

- Energy market: Gas is expected to have an important role to play as part of the future energy mix, but will be dependent on price. A reform of the NEM may also be required to cater for a scenario of a significant penetration of RE.

The findings of each participant have been analysed and the observed patterns have been compared against the expected patterns for each participant.

4.1   PARTICIPANT 1: DIRECTOR OF POWER

4.1.1   Background

A face to face standardised open ended interview was carried out with the Director of Power (**DP**[18]$_{Eng,Comm}$) for the firm. **DP**[18]$_{Eng,Comm}$ is very knowledgeable across all disciplines of power generation, transmission and distribution, which the firm provides consulting and design services. These include renewable energy (wind, solar, hydro) as well as thermal generation (coal and gas). **DP**[18]$_{Eng,Comm}$ has a focus on energy markets from a commercial perspective.

4.1.2   Energy Policy

When asked about the preference between a FIT versus a RPS policy approach, **DP**[18]$_{Eng,Comm}$ indicated that a price on carbon would be a policy measure that would be a preferred option.

DP[18]Eng,Comm: *'My preference would have been for a price on carbon'*

The rationale behind this was explained to be that a carbon price is a mechanism that penalises users based on their pollution footprint, rather than having a mechanism in place which is a payment for RE. A reliance on subsidies such as green certificates which provide a revenue stream is a less favourable mechanism although it is noted that the outcome in terms of RE development may be the same. The indirect effect of a carbon price on the market is favoured because it acts to penalise polluters rather than pay for RE. No specific preference for a FIT or a RPS was indicated.

In terms of the importance of a federal policy going forward, **DP[18]**Eng,Comm stated that this was very important and that policy should not be at the state level.

DP[18]Eng,Comm: *'I don't believe there should be state policies'*

DP[18]Eng,Comm reasoning is that state policy influences where projects are located (i.e. within each state), and may not be as beneficial to the country considering the location of where the best resources are and that the grid is an interconnected network. **DP[18]**Eng,Comm notes the importance of the philosophy of interconnection between states, and transfer of energy rather than each state acting independently.

DP[18]Eng,Comm sees organisations such as ARENA and CEFC as important to RE development. **DP[18]**Eng,Comm opinion is that ARENA is especially useful for providing grants to projects which lenders are not comfortable financing due to their early stage development. **DP[18]**Eng,Comm states that the CEFC is also important in this regard, although not considered to be required anymore for financing of conventional RE technology (wind, solar PV and hydro).

DP[18]Eng,Comm: *'The way I see those two organisations, it's a breeding ground for fostering innovation, but once the market can deal with the risks, let it run'*

4.1.3   Technical

When asked about grid connection cost as a potential barrier to entry for RE development, **DP[18]**Eng,Comm provided a viewpoint that centred around the market being a key driver for innovation to take place. The scenario was put forward that additional grid costs would trickle through to the consumer, and if these costs became unreasonable to the end consumer, more consumers would be incentivised to go off-grid.

So, in order to retain customers and the value of the network, the market would be pressured to innovate to keep costs at a reasonable level. **DP[18]Eng,Comm** expects that price stability in terms of grid connection would be maintained once greater certainty is achieved in terms of the development and adaption of the grid to a new energy mix.

DP[18]Eng,Comm favours the development of REZ *'to expedite the use of the best natural resource'* and to provide direction for the construction of new interconnectors, which would have a greater value if they consider locations of REZ.

4.1.4   Electricity Market

The question around how RE will be complimented in future is communicated by **DP[18]Eng,Comm** to be dependent on future gas prices which is difficult to predict and depends on the export market, and availability of gas. **DP[18]Eng,Comm** notes that gas can play a key part, although will be dependent on the price and fluctuation thereof. Battery technology could have a part to play depending on its development, with some developments currently assessing a combination of gas and battery technology. **DP[18]Eng,Comm** also notes the development of other fuel sources such as hydrogen, which can be generated with excess power from RE plants, and may have an influence on the energy mix as another generation type in future.

In terms of how the NEM operates, **DP[18]Eng,Comm** communicated that should the scenario presented unfold, that there would be signals that would indicate that the market is not functioning correctly and the Australian Energy Market Commission (AEMC) would then be prompted to make changes if required. The question around the suitability of the NEM as a free market going forward was noted by **DP[18]Eng,Comm** to be an interesting scenario which would require further consideration.

4.1.5   Analysis of **DP[18]Eng,Comm.**

DP has indicated a preference to a carbon pricing scheme as opposed to a policy which effectively provides a subsidy for RE. A preference towards FIT versus RPS is not communicated. This is in contrast with the literature which refers to a carbon tax as a fiscal policy measure generally acting as a supplementary policy (Abdmouleh *et al.*, 2015; Aquila *et al.*, 2017) supporting FIT or RPS.

DP[18]_{Eng,Comm} notes the importance of a federal policy, and communicates a dissatisfaction with the existence of individual state policy. **DP[18]**_{Eng,Comm} view on federal policy aligns with the literature on the key importance of long term policy stability (Byrnes *et al.*, 2013; Abdmouleh *et al.*, 2015; Polzin *et al.*, 2015; Diaz-Rainey and Sise, 2018; Simshauser and Tiernan, 2018). DP provides validation for the statement with regards to state policy, although there is some contrast with the literature which refers to states having managed to continue and boost development of RE at times when federal intervention was lacking (Nelson *et al.*, 2017; Diaz-Rainey and Sise, 2018).

Organisation such as CEFC and ARENA are noted to have a role to play for development of less mature RE, although are not required for assisting mature RE technology. This is contrary to the viewpoint of Byrnes and Brown (2015) who raises concern over the potential for the CEFC to be relinquished in future. A more recent publication confirms the importance of CEFC and ARENA in providing 'a specialist toolbox for the Australian government to encourage further valuable innovation and investment' (Miller, 2018).

DP[18]_{Eng,Comm} raises the point that the market would be pressured to respond to increasing grid costs through innovation, to retain asset value. **DP[18]**_{Eng,Comm} essentially explains what Nelson *et al.* (2017) refers to as a death spiral that may have already started to occur in the NEM, with rising costs of grid infrastructure, prompting more customers to leave and become self-sufficient, resulting in the remaining customers being burdened with a greater proportion of the costs. However, as **DP[18]**_{Eng,Comm} points out, the market would be pressured to resist this death spiral and innovate otherwise be pushed out. **DP[18]**_{Eng,Comm} does not raise specific concern around grid connection costs as a barrier, as the market is expected to adapt through innovation. This contrasts with the view of Klessmann *et al.* (2011), in that grid connection cost is one of the barriers to entry for RE development. The development of REZ is considered by **DP[18]**_{Eng,Comm} to be a viable solution for concentrating RE and reducing grid costs, which aligns with what is suggested in the ISP (AEMO, 2018a). **DP[18]**_{Eng,Comm} also notes the importance of interconnectors, and the need for this to be co-ordinated which aligns with the views of Hua *et al.* (2016). The view taken by **DP[18]**_{Eng,Comm} in that grid related issues could be resolved by the market adjusting accordingly is a more optimistic view than what is highlighted in the literature with regards to grid related barriers to RE development (Klessmann *et al.*, 2011; Byrnes *et al.*, 2013; Fouquet, 2013; Abdmouleh *et al.*, 2015; Eleftheriadis and Anagnostopoulou, 2015; Hua *et al.*, 2016; Nelson, 2018).

Gas as a future generation type is not discounted by **DP18[Eng,Comm]**, as referred to in the ISP (AEMO, 2018a), although **DP18[Eng,Comm]** notes that the future of gas pricing is difficult to predict, which is in line with the literature which states that the generation mix will be dependent on the price of gas in the future (Nelson, 2018). Although the potential for other technologies/fuel sources is highlighted by **DP18[Eng,Comm]**, such as hydrogen. As these technologies are emergent, it is difficult to predict future trends.

DP18[Eng,Comm] does not provide additional insight into whether the NEM will be a satisfactory market for an environment of predominantly RE going forward, but does note that it will be forced to adapt based on signals emanating from the market. **DP18[Eng,Comm]** acknowledges the potential validity of the scenario presented, which would require further consideration but does not raise it as an immediate barrier, as the market is expected to adjust in real time. Therefore, the philosophy of a shift away from a free market that may in future be governed by capacity payments and subsidies is not discounted by **DP18[Eng,Comm]** (Byrnes *et al.*, 2013; Cludius *et al.*, 2014; Nelson *et al.*, 2017; Blazquez *et al.*, 2018; Diaz-Rainey and Sise, 2018).

Table 4-1 summarises the data collected and the analysis conducted in this section in terms of the agreement with the expected patterns identified in the literature.

Table 4-1: Summary of the analysis of data collected (DP18[Eng,Comm])

	Theoretical patterns									
	Energy policy				Technical			Energy market		
	Federal regulatory policy	State regulatory policy	Fiscal policy	Other	Grid access (cost)	Grid stability and REZ	Other	Energy Mix (ISP)	NEM reform	Other
DP – Observed patterns alignment (yes, no, partial, NA)	Yes, federal policy is key. No preference for FIT or RPS.	No, preference for state policy to be discouraged.	Partial, carbon pricing noted as a preferred key policy.	NA	No, not funda-mental, market will adapt.	Yes, REZ considered viable.	NA	Partial, gas dependent on a number of factors including technology.	Partial, NEM will adjust, not considered to be a barrier.	NA

4.2.1   Background

A face to face standardised open ended interview was carried out with the Section Manager of Renewables (**SMR[17]Eng.**) at the firm. **SMR[17]Eng.** has extensive expertise in renewable energy and carbon capture and storage. **SMR[17]Eng.** has a focus on business development, team leadership and providing technical advisory services. It is noted that **SMR[17]Eng.** is also a member of the ARENA panel, and has valuable insight into the RE market and future RE developments in Australia.

4.2.2   Energy Policy

SMR[17]Eng. communicated that a target at state and national level is important but did also note that corporates are currently driving the uptake of RE in Australia. No specific preference for a FIT or RPS at utility scale was communicated. Although it has been seen that the uptake in the domestic market is significant when a FIT is employed. **SMR[17]Eng.** stated that it would be ideal to have a federal policy backing the industry, with the states following suit, setting their own targets.

SMR[17]Eng.: *'Having the whole nation setting a policy and targets would be ideal, and then the states getting on board and setting their own particular targets.'*

The CEFC and ARENA were seen by **SMR[17]Eng.** as important industries in Australia, which have been successful in enabling first mover projects, to get across the line. They provide the momentum and sufficient cost reduction for new technology projects to eventually be able to move ahead independently. **SMR[17]Eng.** noted that ARENA is still very active with achieving significant progress just in the past year.

SMR[17]Eng.: *'although I've actually just been involved in the last nine months, what they've (ARENA) actually achieved has been quite phenomenal, in terms of solar and batteries…'*

In terms of ARENA's focus **SMR[17]Eng.** noted that its quite broad and includes, R&D, technology as well as a systems focus.

4.2.3   Technical

In response to the identified issue around future constraints around grid connection, especially in terms of increasing costs **SMR[17]_{Eng}**. noted that pumped hydro would be fundamental in changing the energy landscape.

SMR[17]_{Eng}.: *'It covers all regions across Australia, so it's about stepping back strategically and looking at where the real renewable energy zones are going to be, and I think it's going to look quite different to our old coal fired power generation areas'*

SMR[17]_{Eng}. outlined a strategy involving widespread pumped hydro across the region, which could assist in strengthening the grid in certain areas which RE could then connect into. This would reduce potential grid costs associated with additional plant requirements.

SMR[17]_{Eng}.: *'If you could match pumped hydro energy storage with those renewable energy zones that would then alleviate extra grid connection costs.'*

SMR[17]_{Eng}. promotes the idea of REZ to be mapped to include pumped hydro in each of these zones, and that these solutions would need to be implemented before RE reached 50 percent penetration, at which point the grid is not expected to be able to function without such measures being in place. **SMR** communicated that investigation is already underway to identify these sites, as the lead time on development and construction of pumped hydro is a number of years.

SMR[17]_{Eng} notes that for REZ to be implemented, co-ordination at a federal level will be required. REZ should be driven by federal government, especially the installation of new interconnectors. This will require a firming up of federal policy. There also needs to be close collaboration between the government and findings of an emergent ISP.

4.2.4   Electricity Market

SMR[17]_{Eng} noted that with the uncertainty around the price of gas, investment decisions are deferred, and it is seen as an investment risk with fluctuating prices which are not able to be locked in over the life of the project. **SMR[17]_{Eng}** notes that the ramp up time for pumped hydro should be quite responsive, so given the solution of multiple pumped hydro plants complimenting wind and solar, gas generation may not be necessary.

SMR[17]Eng acknowledges that the NEM will require changes as Australia moves towards a higher penetration of RE. **SMR[17]Eng** notes that the market has moved so quickly that there are already areas which require revisiting. Questions around the developing market for frequency control have already emerged, and whether this should be a service provided by the NSP rather than the generators. **SMR[17]Eng** agrees that during a period of supply exceeding demand, that wholesale pricing could fall very low with greater penetration of RE. It is noted to be something that requires greater consideration, and would have to be revisited.

4.2.5   Analysis of **SMR[17]Eng.**

In terms of energy policy, **SMR[17]Eng** does not indicate a particular preference to a FIT or RPS, FIT is noted to be a very successful scheme in promoting uptake at the domestic level. Byrnes *et al.* (2013) and Diaz-Rainey and Sise (2018) indicate a preference to FIT to promote RE at the utility scale, which **SMR[17]Eng** does not necessarily indicate. **SMR[17]Eng** advocates a federal policy followed by individual state policy to support development. **SMR[17]Eng** has a view on federal policy that partially aligns with the literature on the key importance of long term policy stability (Byrnes *et al.*, 2013; Abdmouleh *et al.*, 2015; Polzin *et al.*, 2015; Diaz-Rainey and Sise, 2018; Simshauser and Tiernan, 2018). **SMR[17]Eng** does note that RE is being driven by corporates, independent of policy support mechanisms. This indicates a lesser importance of state and federal policy, in driving RE directly. **SMR[17]Eng** also notes that policy is required to co-ordinate the development of interconnectors and the success of REZ, which would need to be driven at a federal level, linking back to what is outlined under what emerges in future versions of the ISP. **SMR[17]Eng** views indicate a shift away from the importance of incentivising RE development, to ensuring the infrastructure is in place for continued development. ARENA is noted to be an important government body involved in funding of pumped hydro feasibility project, which is seen to be of vital importance of for future RE development.

A major barrier to RE development is highlighted by **SMR[17]Eng** to be grid instability and associated connection costs at increasing penetrations as RE approaches 50 percent. This aligns with what is outlined in the literature in terms of grid availability and the cost of grid connection being one of the significant barriers to RE development (Klessmann *et al.*, 2011; Byrnes *et al.*, 2013; Fouquet, 2013; Abdmouleh *et al.*, 2015; Eleftheriadis and Anagnostopoulou, 2015; Hua *et al.*, 2016; Nelson, 2018). A potential solution is put forward by **SMR[17]Eng** as distributed pumped hydro, which is noted to have grid forming capability providing system restart ancillary services that conventional generation normally provides (AEMO, 2018a) around which RE could be populated. **SMR[17]Eng** notes that close collaboration will be required at a federal level for development of the REZ and interconnectors, which is also a view expressed by Hua *et al.* (2016).

SMR[17]Eng does not foresee gas playing a pivotal role in the future energy mix, which contradicts what is currently set out under the ISP. The reason for this is the uncertainty surrounding gas prices, this view aligns with concerns raised in the literature surrounding high gas pricing (Nelson, 2018). **SMR[17]Eng** acknowledges that the NEM will require restructuring in future with an increasing penetration of RE, but does not directly respond to the scenario put forward, in that future downward pressure on pool pricing will be a result of increased penetration of zero marginal cost RE (Byrnes *et al.*, 2013; Cludius *et al.*, 2014; Nelson *et al.*, 2017; Blazquez *et al.*, 2018; Diaz-Rainey and Sise, 2018). **SMR[17]Eng** does note the complexity of the question and expects that the scenario may have to be revisited for consideration.

Table 4-2 summarises the data collected and the analysis conducted in this section in terms of the agreement with the expected patterns identified in the literature.

Table 4-2: Summary of the analysis of data collected (SMR[17]Eng)

	Theoretical patterns									
	Energy policy				Technical			Energy market		
	Federal regulatory policy	State regulatory policy	Fiscal policy	Other	Grid access (cost)	Grid stability and REZ	Other	Energy Mix (ISP)	NEM reform	Other
SMR – Observed patterns alignment (yes, no, partial, NA)	Partial, noted to be important. Sector being driven independently. Policy promoting grid development key.	Partial, beneficia but not fundamental going forward.	Yes, ARENA and CEFC have important role to play – develop-ment of pumped hydro.	Policy around grid develop-ment rather than RE is key.	Yes, solutions required prior to 50% RE.	Yes, REZ potential solution to emerge alongside pumped hydro. Need for federal driver.	NA	No, gas pricing uncertainty will discourage investment.	Partial, NEM may need reform. No additional insight.	NEM requires consider-ation around provision of ancillary services.

4.3.1   Background

A face to face standardised open ended interview was carried out with the Engineering Manager of Wind (**EMW[14]Eng.**). **EMW[14]Eng.** has a number of years of experience carrying out wind and solar consulting services for clients in Australia, and has a thorough understanding of the Australian energy market. EMW is responsible for managing the wind advisory and construction services team and providing advisory services in the wind sector.

4.3.2   Energy Policy

EMW[14]Eng. communicated a preference for the carbon price as the most technology neutral approach, to incentivise clean over conventional generation.

EMW[14]Eng.: *'I prefer the idea of the carbon price, not that we have been able to put that into practice'*

In terms of other policy measures, **EMW[14]Eng.** noted that FITs have worked well, and that the RPS has also been a good scheme, implemented in other countries such as the UK. It provides the additional incentive to get the projects built but does not penalise carbon emissions. **EMW[14]Eng.** standpoint is that the aim is to reduce carbon emissions, and therefore generators with high carbon emissions should be penalised, and those with low emissions should benefit. **EMW[14]Eng.** raises the point that the issue with the RPS in Australia was not so much with the policy itself, but with the fact that there was uncertainty surrounding the policy along the way, which stalled RE development. **EMW[14]Eng.** notes the importance of industry certainty in promoting investment and reducing costs.

EMW[14]Eng. points out that the auction process and applying a FIT with contract for difference approach, as seen in state based schemes can be challenging in terms of selecting the correct tariff. Underbidding has caused problems, and can have an impact on project realisation rates, as seen in Brazil.

The importance of the state led RE development schemes is acknowledged.

EMW[14]Eng.: *'It's served as an interim impetus for the industry, especially the ACT auctions that came about at a perfect time when the industry was not doing much because of federal uncertainty…'*

Although the importance of a stable national energy policy commitment is communicated by **EMW[14]Eng.** to be paramount. For industry **EMW[14]Eng.** states that *'national policy is best, it gives the most certainty within the Australian context'*. **EMW[14]Eng.** explains that the disadvantage with state initiatives is that there is a lack of co-ordination between the states which are connection projects into a national grid. In particular, **EMW[14]Eng.** provides an example of the Victorian state initiative which requires the projects to be located within Victoria, which may not be the best option for Australia.

EMW[14]Eng. does not express a view that agencies such as ARENA and CEFC are vital to the continued development of conventional RE development. **EMW[14]Eng.** states that *'ARENA did very well in kick-starting the solar industry…'* which assisted in demonstrating project feasibility to reduce risk to future financiers.

EMW[14]Eng.: *'In terms of conventional wind and solar, ARENA isn't needed anymore'*

EMW[14]Eng. does acknowledge that in terms of developing new technologies, which may be required for a 100 percent RE future, ARENA has a role to play. The same is said for the CEFC, who's mandate includes funding more experimental projects, i.e. projects with battery integration which may not be so feasible, and the institution may have a role to play in investing in these type of projects in future. **EMW[14]Eng.** communicated that local lenders are comfortable with the conventional RE technology and lenders from abroad are also investing.

4.3.3   Technical

In terms of grid constraints impeding RE development, **EMW[14]Eng.** noted that AEMO and NSPs are looking at solutions on how to increase share of RE in the gird. **EMW[14]Eng.** notes that networks have to plan 5 years in advance. The option of developing REZ is put forward by **EMW[14]Eng.** as a current solution being investigated to addressing the escalating grid cost of increasing RE integration. **EMW[14]Eng.** does note that in order for this to be achieved it will be important for community to be on board with the process, as it will entail concentrated areas of RE development, which may be protested.

4.3.4   Electricity Market

When asked about the likelihood of gas being a key part of the future electricity market, **EMW[14]Eng.** communicated that this is possible and is expected to be comprised of mainly peaker plants to provide supply during peak periods or low RE supply. **EMW[14]**Eng. also notes that Victoria and South Australia has a moratorium on gas exploration, and if Australia tapped into their extensive gas reserves, supply would increase driving down eastern gas prices, which currently compete with the export market.

EMW[14]Eng. did not provide a stance on whether the NEM would have to be reformed with the increasing penetration of RE. **EMW[14]**Eng. did acknowledge though, that questions may have to be asked of the NEM and the bidding process in future, given the non-despatchable nature of RE.

4.3.5   Analysis of EMW[14]Eng.

In terms of energy policy, **EMW[14]**Eng. does not indicate a particular preference to a FIT or RPS, noting that both have worked well in promoting RE on a state and federal level. This is in contrast with Byrnes *et al.* (2013) and Diaz-Rainey and Sise (2018) who have communicated a preference for FIT. Although that does not necessarily indicate that the RPS has not been instrumental in the growth of RE, which **EMW[14]**Eng. is advocating. Byrnes and Brown (2015) conclude the same, in that the RPS has been key to promoting RE development in Australia. A carbon tax, which is a fiscal policy measure that has been classified in the literature as a supplementary policy (Abdmouleh *et al.*, 2015; Aquila *et al.*, 2017) supporting FIT or RPS, has been noted by **EMW[14]**Eng. as a preferred policy, which creates a disincentive for conventional energy generation. Carbon pricing is implemented in a number of countries, including other first world countries such as China and various US states (Simshauser and Tiernan, 2018). Although the legislation for a proposed carbon tax was repealed in 2014 (Diaz-Rainey and Sise, 2018), **EMW[14]**Eng. notes the potential for such a policy to be beneficial for reduction of carbon emissions and in turn the development of alternative technologies. **EMW[14]**Eng. notes that other fiscal policy measures (CEFC and ARENA) are seen as less critical for continued development of conventional technology such as wind and solar PV. This is contrary to the viewpoint of Byrnes and Brown (2015) who raises concern over the potential for the CEFC to be relinquished in future. **EMW[14]**Eng.view indicates a potentially more robust market, as less reliance on these institutions is preferred, as grants and loans depend directly on public budget, which in turn makes them more unstable and a less reliable mechanism to spur on the industry (Johnstone *et al.*, 2010). Although the importance of the state targets is acknowledged by**EMW[14]**Eng., the ultimate importance of a stable policy at a federal level is considered to be paramount. This aligns with the literature on the key importance of long term policy stability (Byrnes *et al.*, 2013; Abdmouleh *et al.*, 2015; Polzin *et al.*, 2015; Diaz-Rainey and Sise, 2018; Simshauser and Tiernan, 2018)

In terms of the technical aspects of development, **EMW[14]**[Eng.] acknowledges the necessity for grid expansion in future, which is under AEMO's mandate for planning of the NEM, and is outlined under the ISP (AEMO, 2018a). **EMW[14]**[Eng.] acknowledges that increasing costs could be expected for grid connection in future, as a result of additional equipment being required to maintain stability as RE penetration increases. **EMW[14]**[Eng.] also notes the possibility for a separate market to develop for provision of these services (to provide system stability), as suggested in the ISP (AEMO, 2018a). A potential solution suggested by **EMW[14]**[Eng.] is the development of renewable energy zones to share capital costs of grid infrastructure, such REZ's are described under the ISP (AEMO, 2018a). Specific mention is made to the importance that community influence may have on the success of this solution, as a large area would be subject to concentrated development which may be opposed by local community. Nelson *et al.* (2017) refers to a death spiral that may have already started to occur in the NEM, with rising costs of grid infrastructure, prompting more customers to leave and become self-sufficient, resulting in the remaining customers being burdened with a greater proportion of the costs. In contrast, **EMW[14]**[Eng.] does not raise significant concern over future grid access or pricing for future developments, which is AEMO's responsibility to plan for. **EMW[14]**[Eng.] provides a more optimistic outlook than what is outlined in the literature, in that grid related issues could be a significant barrier to RE development (Klessmann *et al.*, 2011; Byrnes *et al.*, 2013; Fouquet, 2013; Abdmouleh *et al.*, 2015; Eleftheriadis and Anagnostopoulou, 2015; Hua *et al.*, 2016; Nelson, 2018).

In terms of the energy market, future gas peaker plants is noted to be a possibility in future, which aligns with the ISP's determination of increased gas penetration into the energy mix (AEMO, 2018a). **EMW[14]**[Eng.] does not make any specific comment on the reform of the NEM in terms of capacity payments, or subsidies to account for future downward pressure on pool pricing as a result of increased penetration of zero marginal cost RE (Byrnes *et al.*, 2013; Cludius *et al.*, 2014; Nelson *et al.*, 2017; Blazquez *et al.*, 2018; Diaz-Rainey and Sise, 2018). **EMW[14]**[Eng.] does note the complexity of the question and that it may well be the case, but requires further thought.

Table 4-3 summarises the data collected and the analysis conducted in this section in terms of the agreement with the expected patterns identified in the literature.

Table 4-3: Summary of the analysis of data collected (EMW[14]Eng.)

| | Theoretical patterns | | | | | | | | | |
| | Energy policy | | | | Technical | | | Energy market | | |
	Federal regulatory policy	State regulatory policy	Fiscal policy	Other	Grid access (cost)	Grid stability and REZ	Other	Energy Mix (ISP)	NEM reform	Othe
EMW – Observed patterns alignment (yes, no, partial, NA)	Yes, federal policy is key. No preference for FIT or RPS.	Yes, seen as valuable.	Partial, carbon pricing noted as a preferred key policy.	NA	No, not funda-mental, market may adapt.	Partial, REZ could be a solution, stakeholder engage-ment will be key.	NA	Yes, gas considered likely for peaker plants.	Partial, noted as potential require-ment but no further insight.	NA

4.4 PARTICIPANT 4: ENGINEERING MANAGER: SOLAR

4.4.1 Background

A face to face standardised open ended interview was carried out with the Engineering Manager of Solar (EMS[12]Eng.) at the firm. EMS[12]Eng. has previous experience in hybrid generation technologies, and has a number of years of experience in the solar (PV and thermal) sector, with a thorough understanding of the Australian energy market. EMS is responsible for managing the solar advisory and construction services team and providing advisory services in the solar sector.

4.4.2 Energy Policy

EMS[12]Eng. notes that RPS has been quite effective. EMS[12]Eng. states that the development of RE has been occurring at a time of decreasing costs for wind and solar and at the same time of ageing coal fired plants requiring replacement. EMS[12]Eng. makes the point that these factors have driven development of RE as well, although having a policy in place has helped make a clear decision for the replacement of ageing conventional plant with RE.

A federal policy is communicated by **EMS^{12}**Eng. to be preferable from a consistency point of view. Although it is noted that policy is not as fundamental anymore as a requirement for incentivising RE development. **EMS^{12}**Eng. holds the view that it is good that states are picking up the gap in federal policy.

EMS^{12}Eng.: *'Government policy is becoming less and less of a driver for this sector.'*

EMS^{12}Eng. points out that corporate PPAs are being developed which can compete independently as well as the case for merchant plan. Although **EMS^{12}**Eng. expects RE development to become more complicated in terms of grid capacity when penetration gets to roughly 40-50 percent. Policy should therefore be focussed around planning. Economic incentives for wind and solar are not expected to be required. Grid management is expected to be key to successful integration beyond this level of penetration. **EMS^{12}**Eng. suggests that policy direction should be linked with ISP.

EMS^{12}Eng.: *'It (the ISP) is a good foundation that still needs a bit of works, but could really become the definitive basis for policy.'*

EMS^{12}Eng. considers both ARENA and CEFC to be relevant, although not particularly for funding mature technologies. In terms of their importance, CEFC do provide investment in areas where commercial lenders would be too conservative. **EMS^{12}**Eng. does note the success of ARENA in helping to bring down the cost curve for large scale solar previously. **EMS^{12}**Eng. also points to the importance of the funding provided for PV R&D at University of NSW. This has had a potentially significant global impact on module efficiency. ARENA is also funding a lot of pumped hydro feasibility stage assessments. Although enough developers are now able to fund their own feasibility assessments. **EMS^{12}**Eng. considers ARENA to still have a role, but is less clear on their focus. **EMS^{12}**Eng. considers that it would be beneficial for further R&D into other technologies like hydrogen economy and fundamental battery technology.

4.4.3   Technical

EMS^{12}Eng. expects grid connection to be a barrier to projects due to a lack of capacity and additional costs associated with ancillary equipment required to maintain grid stability. A solution is not definitively suggested by **EMS^{12}**Eng.. There is already noted to be a lack of co-ordination between projects where collaboration could reduce costs. A potential solution suggested by **EMS^{12}**Eng. is for the co-ordination of grid infrastructure projects, which could be championed through federal policy which links with the ISP, as discussed previously. **EMS[12]Eng.** communicated that further works is required on the ISP, particularly with regards to validating the marginal cost curve for utility scale storage, based on actual data. **EMS[12]Eng.** suggests that the way forward may be that the NSP's lead the implementation of grid strengthening equipment with a cost recovery mechanism. If this is done efficiently, it is expected that projects could still proceed and the additional cost would not be a showstopper.

EMS[12]Eng.: *'Proponents can cover it if it's made as efficient as possible. How that allocation is done is a tricky matter. If generators are the cause of the system strength issue, I don't think it's unreasonable that they pay.'*

EMS[12]Eng. makes the point that REZ is not a very well defined concept. **EMS[12]Eng.** notes that the approach seems rational to develop projects in clusters around areas of the grid which have capacity. The key challenge that **EMS[12]Eng.** predicts for REZ will be from a stakeholder perspective, if you are saturating areas with RE. **EMS[12]Eng.** notes that the theory of REZ is untested, although areas such as Canberra to Goulbourn could be likened to a REZ type of scenario with numerous wind farms in the area. **EMS[12]Eng.** makes the point that there is an issue with disgruntles locals in the area. There is subsequently an increased planning risk, should the region becomes upset with the barrage of projects. Victoria is also more complicated because of small landholdings. **EMS[12]Eng.** points out that planning policy is not linked to national electricity policy. Solar PV could also be problematic, as there is a land use sterilisation issue. **EMS[12]Eng.** also notes an issue referred to as the 'heat island effect' which is a debated topic, and is something that is currently being raised during the planning process for some large PV plants. There are queries around local temperature increases as a result of the panels modifying the optical properties of the land. **EMS[12]Eng.** notes that this is an example of how planning issues could stall development of concentrated RE development.

4.4.4 Electricity Market

EMS[12]Eng. expects that gas generation does have a role in future, although would likely be minimum use, with maximum benefit. It would make a good compliment to wind, solar and pumped hydro.

EMS[12]Eng. communicated that the way that the NEM currently operates may become redundant, although the scenario put forward had not been something that had been given much thought previously and a definitive answer would need more consideration. **EMS[12]Eng.** does note though that the scenario suggested does hint that some sort of capacity payment could potentially have some merit.

4.4.5   Analysis of EMS[12]Eng.

EMS[12]Eng. has a view that a federal policy is preferred for consistency, and that the state policies did assist in promoting RE development. **EMS[12]Eng.** has a view on federal policy that partially aligns with the literature on the key importance of long term policy stability (Byrnes *et al.*, 2013; Abdmouleh *et al.*, 2015; Polzin *et al.*, 2015; Diaz-Rainey and Sise, 2018; Simshauser and Tiernan, 2018). **EMS[12]Eng.** also corroborates the literature in that states have managed to continue to boost development of RE at times when federal intervention was lacking (Nelson *et al.*, 2017; Diaz-Rainey and Sise, 2018). **EMS[12]Eng.** does put forward an alternative view to the literature going forward though, with regards to type of policy to be implemented. The need for a FIT or RPS type scheme is not seen as fundamental, as the cost cure for mature RE has reduced to a point where RE is competitive without the need for incentives, and ageing conventional plant requires replacement. **EMS[12]Eng.** holds the view that policy could have an important role in driving grid development in terms of what is proposed in the ISP.

EMS[12]Eng. points out the significant risk of grid constraints becoming a barrier once penetration of RE has increased to 40-50 percent, and the possible need for federal support. This aligns with what is outlined in the literature in terms of grid availability and the cost of grid connection being one of the significant barriers to RE development (Klessmann *et al.*, 2011; Byrnes *et al.*, 2013; Fouquet, 2013; Abdmouleh *et al.*, 2015; Eleftheriadis and Anagnostopoulou, 2015; Hua *et al.*, 2016; Nelson, 2018). **EMS[12]Eng.** commends with what is outlined under the ISP, but notes that further work is required particularly on the marginal cost of large scale storage. **EMS[12]Eng.** notes that the concept of REZ outlined under the ISP (AEMO, 2018a) is untested and there are like y to be hurdles to be overcome, particularly with regards to stakeholder engagement and projects achieving planning approval.

The future energy mix could see gas as being a compliment to wind, solar and pumped hydro according to **EMS[12]Eng.**. This aligns with what is set out in the ISP (AEMO, 2018a) and the views expressed by Nelson (2018), although noting that this will be dependent on future gas prices. **EMS[12]Eng.** did not provide a direct response with regards to the functioning of the NEM going forward, stating that further consideration is required, but did not discount the literature in terms of a potential reform required of market operations (Byrnes *et al.*, 2013; Cludius *et al.*, 2014; Nelson *et al.*, 2017; Blazquez *et al.*, 2018; Diaz-Rainey and Sise, 2018).

Table 4-4 summarises the data collected and the analysis conducted in this section in terms of the agreement with the expected patterns identified in the literature.

Table 4-4: Summary of the analysis of data collected (EMS[12]Eng.)

	Theoretical patterns									
	Energy policy				Technical			Energy market		
	Federal regulatory policy	State regulatory policy	Fiscal policy	Other	Grid access (cost)	Grid stability and REZ	Other	Energy Mix (ISP)	NEM reform	Other
EMS – Observed patterns alignment (yes, no, partial, NA)	Partial, FIT or RPS not fundamental going forward. Consistent policy is important (linked to grid).	Partial, state policy has been valuable, but not fundamental going forward.	Yes, CEFC and ARENA do have a role to play.	A policy tying back to the ISP was noted to be fundam-ental.	Yes, grid capacity and cost noted to be potential future barriers.	Partial, REZ noted to be theoretical, not tested yet.	Con-straints on solar due to land use sterili-sation and 'heat island effect'	Yes, mix potentially viable, but ISP needs further work.	Partial, noted as potential require-ment but no further insight. further insight.	NA

4.5.1   Background

A face to face standardised open ended interview was carried out with the Engineering Manager of New Energy (**EMNE[14]Eng.**) at the firm. **EMNE[14]Eng.** has extensive expertise in renewable energy in particular solar and photovoltaics. **EMNE[14]Eng** holds a PhD in Photovoltaic Engineering. **EMNE[14]Eng** has a focus on future energy solutions and developments. Current projects being undertaken include those related to hybrid off-grid systems, the hydrogen economy, and various forms of energy storage.

4.5.2   Energy Policy

EMNE[14]**Eng** showed no specific preference for a FIT or RPS at utility scale, but noted that both contributed to significant time pressures to realise a project and take advantage of the scheme. This has resulted in flooding of the market creating boom and bust cycles, typical of the Australian market. Uncertainty around the permanency of the policies which have been seen to be repealed with changes in government administration has also contributed to the boom and bust cycles.

EMNE[14]**Eng**: *'We saw a lot more pressure and time pressure, and everybody competing at the same time which creates that boom and bust cycle which I think is very typical of Australia and how things have worked with feed in tariffs.'*

EMNE[14]**Eng** noted that a federal policy is more sensible due to the electricity grid being spread across all states in the NEM. Although it is noted that RE is now at a point that the mature technologies are cost effective and cheaper than conventional technologies, as such less focus is expected to be required on certificate schemes and feed in tariffs. Further focus on the grid development and national electricity rules is required to enable fair entry into the market for all industry participants. This will be key to enable further RE development as grid connection is becoming one of the major barriers to further integration of RE.

EMNE[14]**Eng**: *'It's obvious now that the grid needs a lot of work to accommodate RE in those weak areas at the moment, because often those are the best resources.'*

The CEFC and ARENA were seen by **EMNE**[14]**Eng.** as important in supporting enabling technologies for RE, such as energy storage and the hydrogen market. Mature technologies such as wind and solar PV were not expected to require the support of these agencies, other than those projects that may be located in weak parts of the grid and would therefore be exposed to greater risk associated with grid connection.

EMNE[14]**Eng**: *'I'd see its focus moving towards the enabling technologies for RE, storage particularly and things like hydrogen to help with building up those other parts of the market that aren't yet cost effective in their own right'.*

4.5.3   Technical

EMNE[14]**Eng** acknowledges that grid connection costs are becoming an issue with the number of projects entering the market. With more and more projects being connected to the grid, individual projects are having to fund additional plant to comply with electricity rules, with such costs not necessarily having been accounted for initially.

EMNE[14]Eng makes the point that increasing connection costs is becoming a significant barrier to development of RE projects: *'I think it's becoming quite a barrier and making it not as attractive to do renewables in Australia.'*

The current system may be unfair and that the costs should perhaps be spread over a number of projects or be funded partly by government, rather than one project having to bear the costs for additional plant such as a synchronous condenser.

EMNE[14]Eng agrees that the idea of REZ may be a logical solution to enable the remote integration of RE. Although notes that the current marginal loss factors (MLF) scheme may be a disincentive for projects connecting away from a load centre where numerous other projects are located. However, if the grid to these REZ was upgraded sufficiently then this would reduce the impact of MLFs affecting profits.

4.5.4   Electricity Market

EMNE[14]Eng makes the point that Australia has been very conservative in the approach to renewables, and with the retirement of coal fired power stations, the use of conventional gas fired generation may be considered the easy solution as a transition to a greater share of RE. The use of a range of technologies to achieve the transition to RE such as pumped hydro and energy storage could be explored to achieve this transition, without the need to employ conventional gas fired generation specifically. Improvement in demand side management could also have a role to play.

EMNE[14]Eng: *'There's been huge technological advancements and it may not be only on the generation side but also the demand side that we can do something too'.*

When asked about the current structure of the NEM and whether reform may be necessary to accommodate a future energy mix consisting predominately of RE, **EMNE[14]Eng** expects that the NEM would definitely need to be reformed and the current system is based on a conventional model, which doesn't take into account the changing nature of the market. Energy storage may however be able to align better with the current market mechanisms in being able to store energy at times of low pricing and then generate at peak times to take advantage of higher pricing.

4.5.5 Analysis of EMNE[14]Eng.

In terms of energy policy, **EMNE[14]Eng** does not indicate a particular preference to a FIT or RPS, and notes that both, coupled with policy uncertainty have contributed to the boom and bust cycles seen in Australia. Byrnes *et al.* (2013) and Diaz-Rainey and Sise (2018) indicate a preference to FIT to promote RE at the utility scale. In contrast **EMNE[14]Eng** does not advocate either policy as being required in the market given the competitiveness of RE, and points out that focus is required around policy implementation to support grid development and around the electricity rules for fair application to all market participants. **EMNE[14]Eng** view on federal policy aligns with the literature on the key importance of long term policy stability (Byrnes *et al.*, 2013; Abdmouleh *et al.*, 2015; Polzin *et al.*, 2015; Diaz-Rainey and Sise, 2018; Simshauser and Tiernan, 2018).

A major barrier to RE development is highlighted by **EMNE[14]Eng** to be grid connection costs borne by developers especially as penetrations of RE increases. This aligns with what is outlined in the literature in terms of grid availability and the cost of grid connection being one of the significant barriers to RE development (Klessmann *et al.*, 2011; Byrnes *et al.*, 2013; Fouquet, 2013; Abdmouleh *et al.*, 2015; Eleftheriadis and Anagnostopculou, 2015; Hua *et al.*, 2016; Nelson, 2018). **EMNE[14]Eng** suggests a change in status quo around allocation of costs is required, for instance distribution of network strengthening costs amongst multiple projects or the introduction of government support. The introduction of REZ as a potential solution as stated under the ISP is considered by **EMNE[14]Eng** to be a plausible plan, however consideration will have to be given to ensure sufficient grid development to ensure projects are not penalised by low MLFs.

EMNE[14]Eng would like to see other technologies employed in the future energy mix other than conventional gas fired generation, which is seen as the conservative option. Although the ISP defines gas as being a key part of the mix, **EMNE[14]Eng** believes that other technological advancements in energy storage including hydrogen and pumped hydro have a role to play in replacing gas for the most part. **EMNE[14]Eng** confirms that the NEM is based on a system that is and will become outdated, and will require an overhaul in order to operate effectively. **EMNE[14]Eng** view aligns with the scenario suggested in the literature, in that future downward pressure on pool pricing will be a result of increased penetration of zero marginal cost RE (Byrnes *et al.*, 2013; Cludius *et al.*, 2014; Nelson *et al.*, 2017; Blazquez *et al.*, 2018; Diaz-Rainey and Sise, 2018).

Table 4-5 summarises the data collected and the analysis conducted in this section in terms of the agreement with the expected patterns identified in the literature.

Table 4-5: Summary of the analysis of data collected (EMNE[14]Eng)

	Theoretical patterns									
	Energy policy				Technical			Energy market		
	Federal regulatory policy	State regulatory policy	Fiscal policy	Other	Grid access (cost)	Grid stability and REZ	Other	Energy Mix (ISP)	NEM reform	Oth
EMNE[14]Eng – Observed patterns alignment (yes, no, partial, NA)	Partial, consistent policy noted to be important, although less importance of traditional RE policy support mechanisms. Policy promoting grid development and consistent electricity rules key.	No, focus should be at a federal level and priority to grid development in the NEM.	Yes, ARENA and CEFC have important role to play – development of RE enabling technologies such as storage.	Policy around grid development rather than RE is key.	Yes, developers are seeing grid costs as a major constraint to project development.	Yes, REZ seen as a potential solution, although focus required on grid development to avoid MLFs becoming a disincentive for development.	NA	No, conventional gas generation not considered ideal. Pressure on the mix to include new storage technology to compliment RE is expected.	Yes, NEM reform is expected to be required to adapt to the functioning of a new market.	NA

4.6.1   Background

A face to face standardised open ended interview was carried out with the Engineering Manager: Electrical **(EME[14]Eng)**. **EME[14]Eng** is a senior control systems and electrical engineer, who has been involved in project delivery for a range of project types including thermal (coal, gas turbine, diesel), wind and photovoltaic generating stations, water treatment, and industrial resources. He is an experienced engineer in the fields of gas turbine, waste heat boiler and auxiliary plant control systems. EME[14]Eng has a good appreciation of the NEM and what the potential barriers might be to further integration of RE, which is why EME[14]Eng was se ected as a participant.

4.6.2   Energy Policy

EME[14]Eng did not advocate any specific RE policy and noted that the FIT that was introduced at state levels was intended to jumpstart the industry to spur on the industry in terms of investment, skills training and manufacturing, and has served its purpose. Implementing such a policy for further incentivising RE growth was not seen as a viable solution.

EME[14]Eng: *'To continue to drive investments, the focus should be on altering that which might get us to the alt 1A, because I think it's just going to cause more problems just continuing with the feed in tariff mechanism.'*

EME[14]Eng suggests that the focus should rather be on how the development of the grid can be supported, such as providing additional frequency inverters and ensuring grid stability. EME[14]Eng points out that focussing on further incentivising the development of wind and solar will be fruitless unless the grid can support an energy mix that is being proposed, dominated by RE sources.

A point made specifically by EME[14]Eng is the importance of the introduction of a policy on the use of gas in the domestic market in the NEM. The local sale price of gas on the east coast is effectively tied to the export market making it non-competitive in the NEM. Unlike in Western Australia, where they have a reserve for domestic use and comestic sale prices are much lower. Introducing a policy similar to that which exists in Western Australia, would incentivise development of gas plants and aid in the transition that is occurring with the decommissioning of coal plants.

EME[14]Eng : *'What I think will be beneficial to the transition will be a gas policy which means that domestic gas supply has to be a certain percentage of overall gas production'*

In terms of the importance of a national energy policy, EME[14]Eng stated that a national policy is important although shouldn't maintain the existing financial incentives, such as focussing on incentivising wind and solar development. The focus should be on the entire system and what adaptions will be required along the way to achieve the intended energy mix.

EME[14]_{Eng}: *'Financial incentives need to change so that you're improving the overall system, not just supporting the installation of one element, it's a complex mix, it's a complex system.'*

EME[14]_{Eng} view is that ARENA and CEFC do still have a role to play, although should not be focussed on mature technologies. **EME[14]**_{Eng} notes that they have supported these in the past and achieved their objectives, which have now changed.

EME[14]_{Eng}: *'I think ARENA should be focussed on the minority or the emerging technologies, rather than now what is main stream solar and wind.'*

EME[14]_{Eng} makes the point that ARENA should be focussed on encouraging research, and that Australia doesn't spend as much on research compared to other developed nations. ARENA should be deciding on where Australia should be allocating funding for research. **EME[14]**_{Eng} notes that there is a lot of theoretical research being done, but this has not translated into practical solutions that can be implemented as such. Research should be focussing on the transition between this very theoretical research to testing the practicality of such research.

4.6.3   Technical

The idea that connection costs may become a barrier to entry for projects will depend on how the development of the grid is managed, according to **EME[14]**_{Eng}. It is certainly an area that requires careful management according to **EME[14]**_{Eng} and is where policy should be in place to guide the development and allocate responsibility in terms of additional plant requirements to ensure system strength.

EME[14]_{Eng} notes that NSPs have the responsibility for maintaining system strength and have pushed on this responsibility to the connecting parties. This is not unreasonable, however expectations from the grid operators to meet the full requirements is onerous. On the other hand, in an area which might have concentrated levels of RE such as a REZ, the environment should be conducive to incentivise generators to connect, such as the grid being capable for these generators to connect, without the need for extensive upgrades required to be borne by connecting parties.

EME[14]Eng view is that a collaborative approach is needed. The generators are not necessarily the best placed to be installing this equ pment either. Grid operators could step in and provide additional plant such as synchronous condensers in central areas which control fault current levels for the grid in the area to allow protection systems to work. Pushing the responsibility onto individual projects does not always benefit other projects in the region, should the project be placed offline for instance, this would impact other projects ability to generate.

EME[14]Eng: *'More forward planning on who's best to provide certain services and activities should be on the table'*

EME[14]Eng states that the intended outcome and benefits to the local community that REZ purports to achieve may not be realised. This is due to the current process of connection, in that initial parties that connect have an advantage over subsequent generators that connect, which may incur higher connection costs to ensure system strength is maintained. This creates a scenario where everyone is rushing and competing to have their project built first, creating a boom and bust scenario.

EME[14]Eng: *'REZ is a novel and a grand idea, but without putting the systems in place to enable it to occur I don't think it's going to provide the outcome intended.'*

EME[14]Eng concludes that without follow on policy that puts the cost on upgrading the network on effectively multiple projects that could share the overall costs of the grid upgrade, as well as the grid operator ensuring that these projects occur to strengthen the grid, the idea of REZ will remain just an idea.

4.6.4   Electricity Market

EME[14]Eng view is that the scenario put forward in the ISP in terms of energy mix, in particular the requirement for gas fired power plants in the mix is very likely. **EME[14]Eng** notes that the more solar farms that come online, the increased pressure that is being placed on coal plants that could result in being shut down. This is due to the financial viability of these plants given the low demand for electricity during the middle of the day. Gas will therefore play a crucial part of that future energy mix. **EME[14]Eng** makes the point that there isn't enough capacity to cover the peak generation periods with the future loss in generation with coal plants coming offline, and gas plants can't be built fast enough.

EME[14]~Eng~: '*I think gas will play a crucial part in that energy mix, but the issue is there isn't enough storage technology implemented at the required scale to offset the coal fleet coming off line, and there is quite a delay in constructing gas plants.*'

EME[14]~Eng~ did not have a defined view on whether a complete market reform would be required, and is something that would require more thought to be able to opine on. **EME[14]**~Eng~ did provide the view that ancillary services would need to be increased in value, so market reform may be required in those aspects. Ancillary services such as black start capability and frequency response services (6 second and 30 second services) are critical in supporting the functioning of the grid and are being lost with the exit of coal plants which traditionally provide these services. Provision of these services could be a significant future revenue stream for projects, and that could potentially extend the viability of coal plants. **EME[14]**~Eng~ noted a technical issue in the limited ability for the new technologies such as wind, solar or gas for that matter to be able to provide black start capability.

EME[14]~Eng~: '*The grid has been designed around a large coal fired plant being your black start.*'

EME[14]~Eng~ concludes that the shift to a mix dominated by RE is a long programme that is required. Consideration of all the engineering aspects is required and planning should determine what drivers are required to be implemented along the way to get to the end goal.

4.6.5   Analysis of **EME[14]**~Eng~.

In terms of energy policy, **EME[14]**~Eng~ does not advocate either FIT or RPS. The traditional policy approach is seen as being irrelevant to the status of development in Australia, with other policy mechanisms around grid development and policy around gas expected to be required. This is in contrast with the focus in the literature which advocates incentivising RE directly, such as Byrnes *et al.* (2013) and Diaz-Rainey and Sise (2018) who have communicated a preference for a FIT system. **EME[14]**~Eng~ views do however align with the literature in that energy policy needs to address the mechanisms required to promote an optimal energy mix, and how the RET should be structured (Nelson *et al.*, 2015). **EME[14]**~Eng~ views align quite closely with Nelson *et al.* (2017) who argues that a regulatory closure policy which would be based on generator age, could potentially address political concerns around pricing volatility. Upon the planned withdrawal of firm capacity, the suitable replacement generation would need to be incorporated into the market. EME[14]Eng is promoting gas as that replacement generation, and a policy around this to enable such replacement generation to be built.

EME[14]Eng notes that other fiscal policy measures (CEFC and ARENA) have served their purpose for conventional technology such as wind and solar PV, and are not required to support this mature technology any longer. This is contrary to the viewpoint of Byrnes and Brown (2015) who raises concern over the potential for the CEFC, as a supporter of utility scale projects, to be relinquished in future. ARENA's importance is noted however, for its focus of technological research.

EME[14]Eng notes the importance of policy in driving grid development and not what is traditionally referred to in the literature in terms of the key importance of long term policy stability, such as FIT or RPS incentive schemes (Byrnes *et al.*, 2013; Abdmouleh *et al.*, 2015; Polzin *et al.*, 2015; Diaz-Rainey and Sise, 2018; Simshauser and Tiernan, 2018)

In terms of the technical aspects of development, EME[14]Eng notes that collaboration around connection costs will be required between the grid operator and connecting parties. EME[14]Eng does not explicitly state that grid development and connection costs will be a critical barrier, but notes the requirement for planning and policy to support this increased connection of the number of RE projects. This generally aligns with the viewpoints outlined in the literature in terms of the potential obstacles faced by RE projects in terms of grid connection (Klessmann *et al.*, 2011; Byrnes *et al.*, 2013; Abdmouleh *et al.*, 2015; Eleftheriadis and Anagnostopoulou, 2015).

EME[14]Eng view on the development of renewable energy zones to share capital costs of grid infrastructure, as described under the ISP (AEMO, 2018a) is sceptical. The current indications do not show a move towards putting the systems in place that will enable the development of a REZ to occur.

In terms of the energy market mix, EME[14]Eng notes that gas is expected to be a critical part of this, which aligns with the ISP's determination of increased gas penetration into the energy mix (AEMO, 2018a).

NEM reform as suggested by EME[14]Eng may be around ancillary services particularly, which could potentially be a significant future revenue stream for projects. An issue which has not been apparent in the reviewed literature, raised by **EME[14]_{Eng}** is around the potential for RE generation being able to provide the full suite of these services. This is a challenge in a system that has been traditionally designed with the operating assumptions around a large coal plant existing. The requirement for ancillary services is however acknowledged in the ISP (AEMO, 2018a). **EME[14]_{Eng}** does not have a view on whether any reform of the NEM in terms of capacity payments, or subsidies to account for future downward pressure on pool pricing may be necessary (Byrnes *et al.*, 2013; Cludius *et al.*, 2014; Nelson *et al.*, 2017; Blazquez *et al.*, 2018; Diaz-Rainey and Sise, 2018). **EME[14]_{Eng}** notes that the question is not something that has been considered previously and would require further thought.

Table 4-6 summarises the data collected and the analysis conducted in this section in terms of the agreement with the expected patterns identified in the literature.

Table 4-6: Summary of the analysis of data collected (EME[14]_{Eng.})

	Theoretical patterns									
	Energy policy				Technical			Energy market		
	Federal regulatory policy	State regulatory policy	Fiscal policy	Other	Grid access (cost)	Grid stability and REZ	Other	Energy Mix (ISP)	NEM reform	Oth
EME[14]_{Eng} – Observed patterns alignment (yes, no, partial, NA)	Partial, policy is important, but not traditional policy to incentivise RE.	No. Considered to be irrelevant in current context.	Yes, ARENA does have a role to play.	Policy around domestic gas use raised, and policy around grid development	Partial, requires planning and support in terms of policy.	Partial, REZ is not considered to be a viable option at present, unless key steps put in place for this to be achieved.	The requirement for ancillary services is raised as a technical barrier which will required to address.	Yes, gas considered to be critical as part of future energy mix. Although policy will be required to enable this.	Partial, reform suggested around ancillary services. Complete reform not confirmed.	NA

The analysis of the data collected and observed patterns from each of the interviews has been compared against the theoretical patterns expected, which has emerged from the literature. These include energy policy measures, technical constraints in terms of grid access and associated cost, and NEM structure. The analysis tests the theoretical validity of the study by noting patterns which emerge that do not align with the theoretical patterns (Johnson, 1997).

A summary of the participants' responses, and how these align with the identified patterns in the literature is shown in Table 4-7.

Table 4-7: Comparison of theoretical and observed patterns

Observed pattern alignment per participant	Theoretical patterns									
	Energy policy				Technical			Energy market		
	Federal regulatory policy	State regulatory policy	Fiscal policy	Other	Grid access (cost)	Grid stability and REZ	Other	Energy Mix (ISP)	NEM reform	Other
DP[18]_Eng,Comm	Yes, federal policy is key. No preference for FIT or RPS.	No, preference for state policy to be discouraged.	Partial, carbon pricing noted as a preferred key policy.	NA	No, not funda-mental, market may adapt.	Yes, REZ considered viable.	NA	Partial, gas dependent on a number of factors including technology.	Partial, NEM will adjust, not considered to be a barrier.	NA
SMR[17]_Eng.	Partial, noted to be important. Sector being driven independently. Policy promoting grid development key.	Partial, beneficial but not fundamental going forward.	Yes, important role to play – develop-ment of pumped hydro.	Policy around grid develop-ment rather than RE is key.	Yes, solutions required prior to 50% RE.	Yes, REZ potential solution to emerge alongside pumped hydro. Need for federal driver.	NA	No, gas pricing uncertainty will discourage investment.	Partial, NEM may need reform. No additional insight.	NEM requires consider-ation around provision of ancillary services.

EMW14_{Eng.}	Yes, federal policy is key. No preference for FIT or RPS.	Yes, seen as valuable.	Partial, carbon pricing noted as a preferred key policy.	NA	No, not fundamental, market may adapt.	Partial, REZ could be a solution, stakeholder engagement will be key.	NA	Yes, gas considered likely for peaker plants.	Partial, noted as potential requirement but no further insight.	NA
EMS12_{Eng.}	Partial, FIT or RPS not fundamental going forward. Consistent policy is important (linked to grid).	Partial, state policy has been valuable, but not fundamental going forward.	Yes, CEFC and ARENA do have a role to play.	A policy tying back to the ISP was noted to be potential future barriers at 40-50% RE.	Yes, grid capacity and cost noted to be potential future barriers at 40-50% RE.	Partial, REZ noted to be theoretical, not tested yet.	Constraints on solar due to land use sterilisation and 'heat island effect'	Yes, mix potentially viable, but ISP needs further work.	Partial, noted as potential requirement but no further insight.	NA
EMNE14_{Eng}	Partial, consistent policy noted to be important, although less importance of traditional RE policy support mechanisms. Policy promoting grid development and consistent electricity rules key.	No, focus should be at a federal level and priority to grid development in the NEM.	Yes, ARENA and CEFC have important role to play – development of RE enabling technologies such as storage.	Policy around grid development rather than RE is key.	Yes, developers are seeing grid costs as a major constraint to project development.	Yes, REZ seen as a potential solution, although focus required on grid development to avoid MLFs becoming a disincentive for development.	NA	No, conventional gas generation not considered ideal. Pressure on the mix to include new storage technology to compliment RE is expected.	Yes, NEM reform is expected to be required to adapt to the functioning of a new market.	NA
EME14_{Eng}	Partial, policy is important, but not traditional policy to incentivise RE.	No. Considered to be irrelevant in current context.	Yes, ARENA does have a role to play.	Policy around domestic gas use raised, and policy around grid development	Partial, requires planning and support in terms of policy.	Partial, REZ is not considered to be a viable option at present, unless key steps put in place for this to be achieved.	The requirement for ancillary services is raised as a technical barrier which will required	Yes, gas considered to be critical as part of future energy mix. Although policy will be required to enable this.	Partial, reform suggested around ancillary services. Complete reform not confirmed.	NA

						to address.			

4.7.1　Energy Policy

All participants agreed on the importance of federal policy, which aligns and reinforces the observed pattern in the literature. However, none of the participants viewed a FIT or RPS type of policy as being fundamental though, indicating that the debate in the literature around advantages and disadvantages of each policy may be superfluous in the Australian context. **EME[14]Eng, EMS[12]Eng, EMNE[14]Eng** and **SMR[17]Eng.,** indicated that policy should rather be focussed on grid management. Policy implementation at a state level was not seen as being required by **EME[14]Eng, DP[18]Eng.** and **EMNE[14]Eng,** who would rather discourage state policy interventions. **EMS[12]Eng.** also notes that state policy is not seen to be a fundamental requirement for RE development going forward. **EME[14]Eng** makes specific reference to a policy around domestic gas generation being required. A carbon tax is noted by both **EMW[14]Eng.** and **DP[18]Eng.** to be the preferred primary policy measure, as opposed to FIT or RPS. The implementation of a price on carbon over a FIT or RPS scheme to promote RE is not a common approach that emerges from the literature.

In terms of fiscal policy, all participants note the benefit that organisations such as CEFC and ARENA have contributed in the industry, but do not see them as being required for mature technology development such as wind, solar PV. **SMR[17]Eng.** put additional emphasis on the importance of ARENA, in providing funding for pumped hydro feasibility studies, which will be a key contribution to RE development going forward. **EMS[12]Eng.** and **EMW[14]Eng.** also noted ARENA's role in development of pumped hydro. **EMNE[14]Eng** mentions that support will be required generally for enabling technologies such as pumped hydro and development of the hydrogen market.

4.7.2　Technical

The relevance of future grid capacity constraints, and cost associated with grid connection as a barrier to future energy development varied between participants. **EME[14]Eng, DP[18]Eng.** and **EMW[14]Eng.** expect the market to adjust accordingly, and AEMO to plan for future expansion, which is currently in process. **EME[14]Eng** notes the importance of planning and policy around allocation of responsibilities for achieving system strength requirements. **EMS[12]Eng., EMNE[14]Eng** and **SMR[17]Eng.** are more explicit in acknowledging the concerns raised in the literature in terms of grid constraints. **EMS[12]Eng.** and **SMR[17]Eng.** raises concern that this could be a barrier once RE reaches the 40-50 percent penetration mark. **EMS[12]Eng.** expects RE development could overcome these barriers with careful grid management. **EMS[12]Eng.** advocates the implementation of wide spread pumped hydro as a solution, also acknowledging grid management at a federal level will be required. Although **EME[14]Eng**, **EMW[14]Eng.** and **DP[18]Eng.** are not raising significant concern, they do refer to market adjustment and future planning which is effectively what **EMNE[14]Eng**, **EMS[12]Eng.** and **SMR[17]Eng.** is referring to, although **EMNE[14]Eng**, **EMS[12]Eng.** and **SMR[17]Eng.** notes that federal backing in the form of regulatory policy would assist in achieving this.

The potential for REZ to be a solution is considered to be a rational approach by most participants. **EME[14]Eng** does not see it as a practical solution that will deliverer expected outcomes without significant interventions to the status quo. **EMW[14]Eng.** notes that this will have to be led at a federal level. It is noted by **EMS[12]Eng.** and **EMW[14]Eng.** that the biggest hurdle will be stakeholder engagement and effectively obtaining planning approval for a scheme such as this to go ahead in practice. **EMNE[14]Eng** notes the importance of adequate grid development in these remote areas, to ensure projects in the REZ are not burdened with low MLFs which would impact on feasibility.

4.7.3   Energy market

The energy mix outlined in the ISP, in terms of gas generation being an integral part is discounted by **SMR[17]Eng.** and **EMNE[14]Eng** who note the role that pumped hydro can play. **SMR[17]Eng.** believes uncertainty in future gas prices will deter investment. **EMNE[14]Eng** believes that the mix could be achieved without the need for conventional gas fired generation being a key component. **DP[18]Eng.** does note that it is difficult to predict future mix with the number of variables, including gas pricing, and emergent technology. **EME[14]Eng**, **EMS[12]Eng.** and **EMW[14]Eng.** do foresee the possibility of gas peaker plants in future to provide a compliment to wind, solar PV and pumped hydro. **EME[14]Eng** in particular, notes their critical importance. Overall there are varying opinions in terms of the role gas fired generation will play in the future mix, the importance of gas fired generation outlined in the ISP may therefore be overstated.

In terms of the operation of the NEM and whether reform might be required, none of the participants discounted the scenario, although further consideration was noted to be required for any sort of definitive answer to be provided. This may suggest that a reform of the NEM as suggested in the literature from an energy only market to the introduction of a capacity based system, may have some merit in the future. A shift in the market to a system where ancillary services could become more influential in terms of revenue generation was noted by **EME^{14}Eng** and **EMW^{14}Eng.**

4.7.4 Chapter Summary

This chapter outlines the study's findings which is based on the data that has been gathered from six participants, through face to face standardised open ended interviews. The theoretical patterns which have emerged from the literature have been detailed and compared to the observed patterns which have emerged during the data analysis. This has been carried out separately for each participant. Finally, a summary of the observed patters from each of the participants is compared against the theoretical patterns. The analysis has tested the theoretical validity of the study by noting patterns which have emerged that do not align with the theoretical patterns. Chapter 5 draws on the findings of this chapter to formulate relevant conclusions.

This chapter outlines how the research objectives were achieved, and concludes on the problem statement, research question and research proposition listed below;

The problem statement to be addressed, as defined in Chapter 1 is as follows:

The continued growth of RE development in Australia is uncertain, given fluctuating investment in the past and the lack of Federal RE policy going forward.

The research question to be answered was:

What are the potential barriers that may need to be addressed and strategies or policies that may be required for continued growth of RE in Australia, based on the perspective of energy specialists at a leading consultancy operating in the built environment in Australia?

The research proposition that has been assessed was:

Based on from the perspective of energy specialists at a leading consultancy operating in the built environment in Australia, the growth of RE in Australia is likely to be contingent upon addressing potential barriers that may emerge and introducing certain policies to incentivise investment.

Finally, the chapter provides an overall conclusion of the study and outlines recommendations for further research.

The research objectives that were stated in Chapter 1 were as follows:

Understand, based on the perspective of energy specialists at a leading consultancy operating in the built environment in Australia;

- *Whether RE development in Australia requires support/market intervention for continued growth.*
- *What the potential barriers are that may hinder continued growth of RE in Australia.*

- *What the strategies and/or policy support mechanisms are that could be implemented to overcome these barriers.*

The research objectives were achieved by carrying out a single case study analysis of an engineering consultancy in Australia, which involved the interviewing of senior management employees in the RE business unit.

The first objective was to determine whether RE development requires support or market interventions for continued growth. This was achieved through questions in the interview focussing on federal and state policy and preference for different policy measures. There was consensus among participants that federal policy was desired for consistency and that it would be beneficial for the development of RE, which aligns with the literature (Aparicio *et al.*, 2012; Buckman *et al.*, 2014; Byrnes and Brown, 2015; Diaz-Rainey and Sise, 2018; Simshauser and Tiernan, 2018). It was noted though, that this was not necessarily a key requirement as the industry has enough momentum for growth to continue in the absence of policy certainty. Furthermore, **EMNE^{14}_{Eng}**, **EME^{14}_{Eng}**, **EMS^{12}_{Eng}** and **SMR^{17}_{Eng}** views that policy around development of the grid to support increased levels of RE was seen as more important than policy to incentivise RE itself. Carbon pricing emerged as a favourable main support mechanism by **EMW^{14}_{Eng}** and **DP^{18}_{Eng,Comm}**, with a policy around local gas generation favoured by **EME^{14}_{Eng}**. The literature (Abdmouleh *et al.*, 2015; Polzin *et al.*, 2015; Aquila *et al.*, 2017; Simshauser and Tiernan, 2018) discusses carbon pricing as a fiscal policy, and FIT and RPS which is discussed in the literature (Aparicio *et al.*, 2012; Buckman *et al.*, 2014; Byrnes and Brown, 2015; Diaz-Rainey and Sise, 2018; Simshauser and Tiernan, 2018) at length, is not seen as fundamental by participants. Fiscal support in terms of funding through ARENA was seen to be important for research and emerging technologies as well as feasibility studies for pumped hydro, which was noted to be key for future energy development.

The second objective was to identify the potential barriers that may hinder continued growth of RE in Australia. The lack of RE policy, as discussed above, was not considered to be a fundamental barrier, which is contrary to the importance of stable policy outlined in the literature (Aparicio *et al.*, 2012; Buckman *et al.*, 2014; Byrnes and Brown, 2015; Diaz-Rainey and Sise, 2018; Simshauser and Tiernan, 2018).

It is noted that the current status of RE development in Australia which has gained significant momentum as discussed by participants, with all participants noting the importance of grid augmentation required, is a different scenario compared to when much of the literature was drafted prior to the boom in development in the past two years. As such policy stability and incentivisation mechanisms may have been key to development previously. Grid connection is highlighted in the literature (Byrnes *et al.*, 2013; Hua *et al.*, 2016; Nelson *et al.*, 2017) to be one of the significant barriers in terms of grid capacity and cost of connection for RE projects. This was raised in the interview with participants to understand whether this may be a barrier in Australia going forward. All participants agreed that measures would be required to enable the grid to continue to operate effectively with a transition to a RE dominated mix. **EME[14]Eng** , **EMW[14]Eng** and **DP[18]Eng,Comm** noted that the market would adapt in such ways to allow for RE to be integrated, acknowledging that grid expansion and measures would have to be taken to allow for continued integration of RE. **SMR[17]Eng** and **EMS[12]Eng** were more pressing in their response, noting that the current electricity system would not be able to support more than 40-50 percent of RE penetration, unless certain measures were taken.

The idea of REZ was considered to be feasible by most participants, with the exception of **EME[14]Eng** who noted that significant measures would be required for the expected outcomes to be delivered. **EMS[12]Eng** and **EMW[14]Eng** noted potential issues with community acceptance and development approval for densely populated areas of RE, being a potential barrier.

The participants did not identify the current structure of the NEM to be a fundamental barrier to RE development, although all participants did acknowledge that market reform may be necessary as higher penetration of RE is achieved. The supplementary technology to support wind and solar that would enable a secure supply of energy in a future RE market was noted by most participants to be dependent on emergent technology as well as variable gas pricing which may deter investment. **EMS[12]Eng EME[14]Eng** specifically mention ancillary services as an important market that would be required with the exit of coal plants for the grid to continue to function.

The third objective was to identify the strategies/support mechanisms that could be implemented to overcome these barriers. An attempt has been made during the interviews to elicit the strategies and solutions to overcoming the barriers as discussed above. None of the participants noted the importance of incentivising RE development itself, as such traditional policy measures such as FIT and RPS were seen as being irrelevant in the current Australian context. **EME[14]**$_{Eng}$, **EMNE[14]**$_{Eng}$, **EMS[12]**$_{Eng}$ and **SMR[17]**$_{Eng}$ highlighted the importance of grid expansion and the strategy under the ISP for integration of RE to be championed at the federal level. **EMS[12]**$_{Eng}$, **EMW[14]**$_{Eng}$ and **SMR[17]**$_{Eng}$ made mention to pumped hydro development, with **SMR[17]**$_{Eng}$ noting the importance that pumped hydro is expected to play in enabling RE development to exceed 40-50 percent penetration. The role that ARENA will play in the development and utilisation of future technologies such as the hydrogen economy and fundamental battery technology will be beneficial, as noted by **SMR[17]**$_{Eng}$. **EMW[14]**$_{Eng}$ and **EME[14]**$_{Eng}$ note a potential reform in the market for ancillary services to incentivise provision of such services which will become more important with the decommissioning of coal plants which traditionally provided such services.

5.3 CONCLUSIONS ON THE RESEARCH QUESTION

The research question was:

What are the potential barriers that may need to be addressed and strategies or policies that may be required for continued growth of RE in Australia, based on the perspective of energy specialists at a leading consultancy operating in the built environment in Australia?

The investigation indicated that although the literature (Aparicio *et al.*, 2012; Buckman *et al.*, 2014; Byrnes and Brown, 2015; Diaz-Rainey and Sise, 2018; Simshauser and Tiernan, 2018) emphasised stable RE policy such as FIT and RPS schemes as one of the key factors to consider for driving RE development, this was not considered to be particularly relevant in the Australian context. Barriers which have been identified and proposed support strategies which are specific to Australia include the following;

1) The transmission system may become a fundamental constraint to the shift to a distributed energy supply. The integration of RE and plans for grid development as outlined under the ISP and the future revisions of the ISP need to be championed at a federal level. Policy support for the development of the national grid to ensure RE penetration is not hindered will be key.

2) RE development is expected to be hindered once penetration reaches 40-50 percent, due to grid constraints, particularly system strength. Development of distributed pumped hydro will be key in supporting system strength in REZ and providing storage for secure energy supply.

3) ARENA has a role to play in funding projects which do not have commercial backing such as the hydrogen economy and fundamental battery technology. This will enable development of new technology which may be key in supporting a market that consists of high penetrations of RE. Funding of pumped hydro feasibility projects will also assist in promoting development of this technology.

4) Community acceptance may be a significant barrier to the development of REZ. Consideration for obtaining development approvals for projects in REZ will be required. Community engagement, and strategic placement of the REZ with consideration to sensitive receptors will be key. The development of the REZ may be required to link with planning policy to ensure that projects are not hindered by appeals.

5) The decommissioning of coal plants also means the exit of the ancillary services required for the grid to function effectively. A reform of the market to incentivise the provision of these services may therefore be required. Consideration of all the engineering aspects such as ancillary services is required and planning should determine what drivers are required to be implemented along the way to achieve the desired outcome.

5.4 CONCLUSIONS ON THE RESEARCH PROPOSITION

The research proposition was:

Based on the perspective of energy specialists at a leading consultancy operating in the built environment in Australia, the growth of RE in Australia is likely to be contingent upon addressing potential barriers that may emerge and introducing certain policies to incentivise investment.

The proposition is partially supported. The participants indicated that certain barriers are expected to emerge with the further integration of RE and would be required to be addressed for the continued development of RE. In particular, two of the participants specifically noted that additional measures would be required once RE penetration reaches levels of 40-50 percent. Grid capability is identified as a barrier to development in the literature (Byrnes *et al.*, 2013; Hua *et al.*, 2016; Nelson *et al.*, 2017) as well as by participants, which would require an adaption to ensure continued RE development. The potential barriers have been identified, which need to be addressed to ensure this adaption.

Growth of RE is not expected to be contingent upon traditional policy support mechanism to incentivise investment. Participants highlighted the importance of grid development being championed at the Federal level, rather than focusing on implementation of RE support policies themselves.

Contrary to the literature (Aparicio *et al.*, 2012; Buckman *et al.*, 2014; Byrnes and Brown, 2015; Diaz-Rainey and Sise, 2018; Simshauser and Tiernan, 2018), conventional RE policy was not identified as being paramount in supporting ongoing development of RE. Although stable energy policy was seen as beneficial for continued development of RE, neither RPS nor FIT was communicated by participants to be key in incentivising further development of RE at a state or federal level. Fiscal support, in terms of funding through agencies such as ARENA was highlighted to have an important role to play going forward. Other fiscal mechanisms, such as a tax on carbon were noted by two participants to be a preferred approach to supporting RE development. Although stable energy policy has been promoted by participants, the potential lack of it was not seen to be detrimental to the industry.

The capability of the grid was identified as a key technical challenge that would have to be overcome to ensure continued RE development at levels approaching 40-50 percent. Grid expansion and interconnection would be required, as well as the development of ancillary services traditionally provided by coal plants. Support at the federal level would greatly benefit these works. REZ which are proposed under the ISP could be a feasible solution to steer grid expansion. Although development approval constraints and community consultation in these areas will be key, and may need backing by planning policy to avoid hurdles and streamline the planning process. Grid strengthening in these areas would also be key in delivering the expected outcomes of the REZ.

The future energy mix that will support an energy market consisting of a majority share of RE will be dependent on emergent technology. The fluctuating gas price may inhibit further development of gas generation, seen as a compliment to RE. A policy around domestic gas generation in the NEM may be key a solution to addressing this. Although gas may have a role to play, it will also depend on options such as battery technology that may advance and become feasible at a larger scale. Pumped hydro is a technology that is expected to be a good compliment to high levels of RE penetration, helping to address key technical constraints associated with high levels of non-synchronous generation in the network, and is expected to be a characteristic of the future energy landscape. Whether the NEM will require reform to support the changing energy landscape was a question that had not been considered in much depth by participants, although it was acknowledged that this may be the case. A shift in the market for ancillary services is expected given the exit of coal plants which previously provided these services, for the grid to function effectively.

5.6 RECOMMENDATIONS FOR FUTURE RESEARCH

The following recommendations for future research have emerged from the investigation:

1) Investigation of the market mechanisms of the NEM and how this is expected to be impacted by increasing penetration of RE.

   The literature identifies a potential reform of the NEM required in future. This is based on the downward pressure on wholesale pricing as a result of increased RE penetration, and the failure of an energy based market as more coal fired power stations are mothballed as a result. The participants had not considered this scenario. Further investigation into the scenario and how the NEM may have to adapt to a changing energy landscape is recommended.

2) The role of pumped hydro in Australia's future energy mix.

   The future energy mix in Australia is noted to be dependent on emergent technology and gas pricing. Pumped hydro has been identified as a key requirement for increased penetration of RE, which could potentially replace the need for gas fired generation as a compliment, and emergent battery technology for storage. An investigation into the feasibility of pumped hydro as the major compliment to a RE dominated market is recommended for further investigation.

3) Development of Renewable Energy Zones

   The development of REZ is a key inclusion in the ISP, which is supported by participants. Although it is noted by one participant that the REZ is theoretical and has not been implemented in Australia. Community consultation is considered to be key, with potential barriers which may emerge during the planning process. It has been recommended that planning policy is considered alongside the identification of the REZ and is adapted to streamline such development. Additional insight may be gained by investigating the development of REZ in countries, identifying the barriers and potential solutions in more detail.

4) Ancillary Services

The grid network has been designed on the traditional assumption of large coal plant generation being present. The exit of coal plants which traditionally provided the ancillary services required for the grid to function puts the functionality of the grid at risk. Further research is recommended into the ancillary services market and how the grid and NEM might have to adapt to function in an environment with limited large scale coal plants in operation.

5) Grid Strengthening

A key requirement noted by all participants is the need for grid interconnection, expansion and strengthening to support RE penetration. In particular, strategies for achieving the required grid capabilities to enable future projects to connect could be investigated beyond those strategies discussed briefly, such as the REZ and individual projects providing additional strengthening equipment.

REFERENCES

Abdmouleh, Z., Alammari, R.A.M. and Gastli, A. (2015) Review of policies encouraging renewable energy integration & best practices. *Renewable and Sustainable Energy Reviews*, **45**, 249-262.

Abdmoulehn, Z., Alammari, R. and Gastli, A. (2015) Review of policies encouraging renewable energy integration & best practices. *Renewable and Sustainable Energy Reviews*, **45**, 249-262.

Abolhosseini, S. and Heshmati, A. (2014) The main support mechanisms to finance renewable energy development. *Renewable and Sustainable Energy Reviews*, **40**, 876-885.

ACCC (2018) Restoring electricity affordability and Australia's competitive advantage - Final Report

AEMO (2018a) Integrated System Plan for the National Electricity Market.

AEMO (2018b) Australian Energy Market Operator. In.

AEMO (2019) Generation Information. In.

Aparicio, N., MacGill, I., Abbad, J.R. and Beltran, H. (2012) Comparison of Wind Energy Support Policy and Electricity Market Design in Europe, the United States, and Australia *IEEE Transactions on Sustainable Energy*, **3**(4), 809-818.

Aquila, G., Pamplona, E.d.O., Queiroz, A.R.d., Rotela Junior, P. and Fonseca, M.N. (2017) An overview of incentive policies for the expansion of renewable energy generation in electricity power systems and the Brazilian experience. *Renewable and Sustainable Energy Reviews*, **70**, 1090-1098.

Arnold, U. and Yildiz, Ö. (2015) Economic risk analysis of decentralized renewable energy infrastructures–A Monte Carlo Simulation approach. *Renewable Energy*, **77**, 227-239.

Baxter, P. and Jack, S. (2008) Qualitative Case Study Methodology: Study Design and Implementation for Novice Researchers. *The Qualitative Report*, **13**(4), 544-559.

Blazquez, J., Fuentes-Bracamontes, R., Bollino, C.A. and Nezamuddin, N. (2018) The renewable energy policy Paradox. *Renewable and Sustainable Energy Reviews*, **82**, 1-5.

Buckman, G., Sibley, J. and Bourne, R. (2014) The large-scale solar feed-in tariff reverse auction in the Australian Capital Territory, Australia. *Energy Policy*, **72**, 14-22.

Byrnes, L. and Brown, C. (2015) Australia's renewable energy policy: the case for intervention. *Munich Personal RePEc Archive*.

Byrnes, L., Brown, C., Foster, J. and Wagner, L. (2013) Australian renewable energy policy: Barriers and challenges. *Renewable Energy*, **60**, 711-721.

Campbell, D.T. (1975) 'Degrees of Freedom' and the Case Study. *Comparative Political Studies*, **8**, 178-193.

Cavaye, A. (1996) Case study research: a multi-faceted research approach for IS. *Information Systems Journal*, **6**(1), 227-242.

CER (2017) *Clean Energy Regulator, Small-scale systems eligible for certificates* [Online].

Clark, A. (2006) *Anonymising research data*. Manchester: ESRC National Centre for Research Methods.

Cleary, M., Horsfall, J. and Hayter, M. (2014) Data collection and sampling in qualitative research: does size matter? . *Journal of advanced nursing*, **70**(3), 473-475.

Cludius, J., Forrest, S. and MacGill, I. (2014) Distributional effects of the Australian Renewable Energy Target (RET) through wholesale and retail electricity price impacts. *Energy Policy*, **71**, 40-51.

Collingridge, D.S. and Gantt, E.E. (2008) The Quality of Qualitative Research. *American Journal of Medical Quality*, **23**(5), 389-395.

Colvin, R., Witt, G.B. and Lacey, J. (2016) How wind became a four-letter word: Lessons for community engagement from a wind energy conflict in King Island, Australia. *Energy Policy*, **98**, 483-494.

Darke, P., Shanks, G. and Broadbent, M. (1998) Succesfully completing case study research: Combining rigour, relevance and pragmatism. . *Information Systems Journal*, **8**(1), 273-289.

Del Rio, P. (2016) *Implementation of Auctions for Renewable Energy Support in Spain: a Case Study*. AURES.

del Río, P. and Linares, P. (2014) Back to the future? Rethinking auctions for renewable electricity support. *Renewable and Sustainable Energy Reviews*, **35**, 42-56.

Denzin, N. and Lincoln, Y. (2011) *The Sage Handbook of Qualitative Research*. Sage Publications Inc.

Diaz-Rainey, I. and Sise, G. (2018) *Green Energy Finance in Australia and New Zealand*. Tokyo:

Eleftheriadis, I.M. and Anagnostopoulou, E.G. (2015) Identifying barriers in the diffusion of renewable energy sources. *Energy Policy*, **80**, 153-164.

Etikan, I., Musa, S. and Alkassim, R. (2016) Comparison of Convenience Sampling and Purposive Sampling. *American Journal of Theoretical and Applied Statistics*, **5**(1), 1-4.

Flyvbjerg, B. (2006) Five misunderstandings of case study research. *Qulaitative Inquiry*, **12**(2), 219-245.

Fouquet, D. (2013) Policy instruments for renewable energy – From a European perspective. *Renewable Energy*, **49**, 15-18.

Geddes, A., Schmidt, T.S. and Steffen, B. (2018) The multiple roles of state investment banks in low-carbon energy finance: An analysis of Australia, the UK and Germany. *Energy Policy*, **115**, 158-170.

Guidolin, M. and Alpcan, T. (2019) Transition to sustainable energy generation in Australia: Interplay between coal, gas and renewables. *Renewable Energy*, **139**, 359-367.

Hirth, L. and Steckel, J.C. (2016) The role of capital costs in decarbonizing the electricity sector. *Environmental Research Letters*(11).

Hua, Y., Oliphant, M. and Hu, E.J. (2016) Development of renewable energy in Australia and China: A comparison of policies and status. *Renewable Energy*, **85**, 1044-1051.

Hyde Kenneth, F. (2000) Recognising deductive processes in qualitative research. *Qualitative Market Research: An International Journal*, **3**(2), 82-90.

IRENA (2018) Renewable capacity statistics 2018.

Jennings, G.R. (2005) *Interviewing: a Focus on Qualitative Techniques. Tourism Research Methods*, Oxfordshire: CABI Publishing.

Johnson, R.B. (1997) Examining the Validity Structure of Qualitative Research. *Education*, **118**(282-292).

Johnstone, N., Haščič, I. and Popp, D. (2010) Renewable Energy Policies and Technological Innovation: Evidence Based on Patent Counts. *Environ Resource Econ*, **45**(1), 133-155.

Kann, S. (2009) Overcoming barriers to wind project finance in Australia. *Energy Policy*, **37**(8), 3139–3148.

Kitzing, L., Mitchell, C. and Morthorst, P.E. (2012) Renewable energy policies in Europe: Converging or diverging? *Energy Policy*, **51**, 192-201.

Kitzing, L., Islam, M., Soysal, E. R., Held, A., Ragwitz, M., Winkler, J., ... Woodman, B. (2016) Recommendations on the role of auctions in a new renewable energy directive.

Klessmann, C., Held, A., Rathmann, M. and Ragwitz, M. (2011) Status and perspectives of renewable energy policy and deployment in the European Union—What is needed to reach the 2020 targets? *Energy Policy*, **39**(12), 7637-7657.

Klessmann, C., Rathmann, M., de Jager, D., Gazzo, A., Resch, G., Busch, S. and Ragwitz, M. (2013) Policy options for reducing the costs of reaching the European renewables target. *Renewable Energy*, **57**, 390-403.

Kothari, C. (2004) *Research Mehtodology: Methods and Techniques*. Second revised edition ed.

Kristinsdottir, A. (2012) *Risks and Decision Making in Development of New Power Plant Projects*. Construction Engineering and Management, Massachusetts Institute of Technoloy.

Lesser, J.A. and Su, X. (2008) Design of an economically efficient feed-in tariff structure for renewable energy development. *Energy Policy*, **36**(3), 981-990.

Lund (2012) Qualitative, quantitative and mixed methods dissertations. In: Lund Research Limited.

MacGill, I., Outhred, H. and Nolles, K. (2006) Some design lessons from market-based greenhouse gas regulation in the restructured Australian electricity industry. *Energy Policy*, **34**, 11-25.

Martin, N. and Rice, J. (2015) Improving Australia's renewable energy project policy and planning: A multiple stakeholder analysis. *Energy Policy*, **84**, 128-141.

Martin, N.J. and Rice, J.L. (2012) Developing Renewable Energy Supply in Queensland, Australia: A study of the barriers, targets, policies and actions. *Renewable Energy*, **44**, 119-127.

Masini, A. and Menichetti, E. (2013) Investment decisions in the renewable energy sector: An Analysis of Non-Financial Drivers. *Technological Forecasting & Social Change*, **80**, 510-524.

McLeod, I. (2017) Integrating Renewables in Australia: Policies, Market Design, and System Operations, *Renewable Energy Integration*, pp. 119-129: Elsevier.

Menanteau, P. (2000) Learning from Variety and Competition Between Technological Options for Generating Photovoltaic Electricity. *Technological Forecasting and Social Change*, **63**(1), 63-80.

Miller, M. (2018) Climate Finance and Financial Markets in Australia: The CEFC and ARENA. *AUSTRALIAN LAW JOURNAL*, **92**(10), 822-829.

Nelson, T. (2018) The future of electricity generation in Australia: A case study of New South Wales. *The Electricity Journal*, **31**(1), 42-50.

Nelson, T., Reid, C. and McNeill, J. (2015) Energy-only markets and renewable energy targets: Complementary policy or policy collision? *Economic Analysis and Policy*, **46**, 25-42.

Nelson, T., Bashir, S., McCracken-Hewson, E. and Pierce, M. (2017) The Changing Nature of the Australian Electricity Industry. *Economic Papers: A journal of applied economics and policy*, **36**(2), 104-120.

Ouyang, X. and Lin, B. (2014) Impacts of increasing renewable energy subsidies and phasing out fossil fuel subsidies in China. *Renewable and Sustainable Energy Reviews*, **37**, 933-942.

Painuly, J.P. (2001) Barriers to renewable energy penetration; a framework for analysis. *Renewable Energy*, **24**(1), 73-89.

Patton, M.Q. (2002) *Qualitative Research and Evaluation Methods*. Sage Publications.

Polzin, F., Migendt, M., Täube, F.A. and von Flotow, P. (2015) Public policy influence on renewable energy investments—A panel data study across OECD countries. *Energy Policy*, **80**, 98-111.

Prest, J. and Soutter, G. (2018) The Future of Australia's Federal Renewable Energy Law.

QueenslandGovernment (2018). Available: https://www.dnrme.qld.gov.au/energy/initiatives/powering-queensland [Accessed August, 2018].

Schaffer, L.M. and Bernauer, T. (2014) Explaining government choices for promoting renewable energy. *Energy Policy*, **68**, 15-27.

Simpson, G. and Clifton, J. (2014a) Consultation, Participation and Policy-Making: Evaluating Australia's Renewable Energy Target. *Australian Journal of Public Administration*, **73**(1), 29-33.

Simpson, G. and Clifton, J. (2014b) Picking winners and policy uncertainty: Stakeholder perceptions of Australia's Renewable Energy Target. *Renewable Energy*, **67**, 128-135.

Simshauser, P. and Tiernan, A. (2018) Climate change policy discontinuity and its effects on Australia's national electricity market. *Australian Journal of Public Administration*, **0**(0).

Stake, R. (1995) *The Art of Case Study Research*. Thousand Oaks, CA: Sage Publications.

Stocks, M., Baldwin, K. and Blakers, A. (2019) Powering ahead: Australia leading the world in renewable energy build rates.

Sun, P. and Nie, P.-y. (2015) A comparative study of feed-in tariff and renewable portfolio standard policy in renewable energy industry. *Renewable Energy*, **74**, 255-262.

Tang, C. and Zhang, F. (2019) Classification, principle and pricing manner of renewable power purchase agreement. *In, IOP Conference Series: Earth and Environmental Science*. IOP Publishing, Vol. 295, 052054.

Tellis, W. (1997) Introduction to Case Study. *The Qualitative Report*, **3**(2).

Trieb, F. (2010) Trans-Mediterranean Interconnection for Concentrating Solar Power. *In, Synergistic Supergrid Conference*, London.

Trochim, W.M.K. (1989) Outcome Pattern Matching and Program Theory. *Evaluation and Program Planning*, **12**, 355-366.

Turner III, D.W. (2010) Qualitative interview design: A practical guide for novice investigators. *The Qualitative Report*, **15**(3), 754-760.

VicGov (2018a) Victoria Renewable Energy Auction Scheme. In.

VicGov (2018b). Available: https://www.energy.vic.gov.au/renewable-energy/victorias-renewable-energy-targets [Accessed August, 2018].

WEC (2016) *World Energy Council, World Energy Resources 2016*. London: World Energy Council.

Wiser, R.H. (1997) Renewable energy finance and project ownership. *Energy Policy*, **25**, 15-27.

WorldBank (2008) REToolkit: A Resource for Renewable Energy Development. In: World Bank.

Wustenhagen, R., Wolsink, M. and Burer, M.J. (2007) Social acceptance of renewable energy innovation: An introduction to the concept. *Energy Policy*, **35**(5), 2683–2691.

Yin, R. (1994) *Case Study Research: Design and Methods* Thousand Oaks, CA: Sage Publications.

Yin, R. (2003) *Case study research: Design and methods (3rd ed.)*. Thousand Oaks, CA: Sage.

APPLICATION FORM

Pleue Note:

Any person planning to undertake research In the Faculty of Engineering and the Built Environment (EBE) at the University of Cape Town is required to complete this fonn **before** collecting or analysing data. The objective of submitting this application prior to embarl<Ing on research Is to ensure that the highest ethical standards in research, conducted under the auspices of the EBE Faculty, are met. Please ensure that you have **read**, and understood the **EBE Ethics In Research Handbook** (available from the UCT EBE, Research Ethica website) prior to complatfng this application form: http://twww.ebe.uct.ac.mbefresearch/ethics1

APPLICANTS DETAILS

Name of principal researcher, student or external appllcant		Ben Campbell
Department		Construction Economics and Management
Preferred email address of applicant		Benjemln.campbell@wap.com
If Student	Your Degree: e.g., MSc, PhD, etc.	MSc.
	Credit Value of Research: e.g., 60/120/180/360 etc.	60
	Name of Supervisor (If supervised):	Saul Nurick
If this Is a research contract, Indicate the source of funding{sponsorship		
Project Iitle		Large-scale grid-connected renewable energy In Australia: Key success fa◆

I hereby undertake to carry out my research In **such** a way that:

* there is no apparent legal objection to the nature or the method of research; and
* the research will not oompromlse staff or students or the other responsibilities of the University;
* the stated objective will be achieved, and the findings wlll have a high degree of validity;
* limitations end altemetlve interpretations will be considered;
* the findings could be subject to pBBr review and publicly avallable; and
* I will comply with the conventions of copyright end avoid any practice that would constitute plagiarism.

SIGNED BY	Full name	Signature	Date
Principal Researcher/ **Student/External applicant**	I Bon Campball	Signature Removed	23 Sep 2018

APPLICATION APPROVED BY	Full name		
8upervlsor (where applicable)	S f\JuR lK Click hero lo enter text.	Signature Removed	
HOD (or delegated nominee) Final authority for all applicants who have answered NO to all questions In Section1; end for all Undergraduate research (Including Honours).	Click here to	Signature Removed	a date.
Chalr: Faculty EIR Committee For applicants other than undergraduata students who have answered YES to any of the above questions.			

Transcription Example: EMW

Q1

Ben: The national policy is that of a Renewable Portfolio Standard with LGCs, compared to a FIT system which is being employed at state level with contract for difference. Should this be the preferred policy?

Yes, it's complicated. Australia has been through many combinations and permutations have been discussed about the way forward. I think from my perspective I prefer the idea of the carbon price, not that we've been able to put that into practice, so you know, maybe it wouldn't work out, but I just see that that should be the most technology neutral and most cost-effective way to encourage renewables into the market.

Yes, clean over conventional rather than ya you know I just find that the feed in tariffs have worked well in the Australian context, so I suppose the renewable energy targets worked well I think and having that price, I think that's been a good scheme. I saw that in practice in the UK as well and I thought that also works quite well – at least provides that incentive to get renewable projects built, but doesn't really penalise the carbon emissions for everybody else, which I feel is a bit unfair which to me comes down to the fundamental question – are saying that we need to reduce our carbon emissions and that's what this whole thing is about and so, therefore, if you have high carbon emissions you should pay for it and if you're reducing them you should benefit from it. It's interesting, like it's interesting with the renewable energy target, like in Australia the only issue we had with it was quite politicised and there were periods along the way where they changed it or they were going to change it and that sort of stalled our industry along the way. They had a good plan for it, but then instead of letting it runs its course, everyone had to have their two cents' worth, so that's caused huge problems in Australia with the stop – start nature of it and I think if they had just let it run its course and given the industry that certainty, we would have won probably filled it by now for sure by now and we would have had benefits in terms of costs because we could have just had a nice ramp down on the cost cove and just given industry that certainty to invest here. The auction process industry – I think there are a few problems with doing auctions because you know, that contract for difference, it's all about driving that lowest cost and because you just pay it, they don't always get the cost right and you've seen issues with like the auction process in Brazil. People have bid way too low and that's caused problems. I suppose the good points about how it's worked here is that it's served as an interim impetus for the industry so especially the ACT auctions came about at a perfect time when the industry was really not doing much because the federal uncertainty with the green price.

Q2

Ben: How important will national energy policy be for continued RE integration, assuming continued state support for RE integration

For industry, national policy is best. It gives the most certainty within the Australian context. The states doing their own thing means that it is much more difficult to navigate in terms of you know, which projects you should go for where. Again it comes back to will the most cost-effective and best for our national grid projects end up being built if each state is trying to do their own thing. The problem as well we've seen with the ACT and Victoria in the state-based scheme, the ACT couldn't have the projects within the ACT because of size reasons so they did end up spread out over Australia so that was sort of helpful and probably encouraged them to go to the best locations within the grid because that brought their price down, but in Victoria they want Victorian projects, which is sensible, given that they are a state-based system – the Victorian tax payers are paying for the tariff, so that's how they're going to sell it. They want Victorian jobs and they want Victorian projects. So whether all the projects should be built in Victoria or not is not ideal, so I think that while I can't hit the federal politicians over the head and make them do anything, it will help us a lot in the next few years, but it will be great to see Australia having a national policy again.

Q3

Ben: Is successful RE development dependent on financing institutions such as the CEFC and ARENA, based on the number of projects that these bodies finance?

I think that ARENA, given that they are more supporting the RE, if they get their funding cut, that's going to be disappointing in terms of pushing our industry forward. I think they did really well in terms of kick-starting the solar industry, by providing grants to the first couple of PB projects that have now rolled on themselves. It's about risk. If the government can show from that first project here's how we can do it and it does work in Australia and then everyone can follow on board. In terms of more conventional like wind and solar, ARENA isn't needed anymore, but maybe there are future technologies they should get involved. That will be the challenge for the industry if they are not there to support that.

Ben: I suppose looking ahead towards 100% renewables, you're going to have other technologies involved – it can't just be wind and solar?

Yes, exactly, it's to get us to that 100% grid. It's not like research and development can stop now, that we can just build wind and solar farms now and it will all be fine and like we'll get to 100%. You're right. That is obviously where their role should be. If they get taken away then I think we'll get to a certain level with wind and solar, but then we may start running into trouble because we've reached the easy wins on the grid. So with the CEFC, they're sort of in the same boat. Their mandate was always more to try and fund more experimental projects. That's why they've gotten into having a battery on everything now because that's where they see promoting it. I don't think batteries are particularly economically viable yet, but they can wear a little bit of that within their ream it, but I think again on the wind and solar front our banks I think are now pretty comfortable with the technology and overseas banks are coming here so I think for the more standard stuff, we're good, but to get to 100% we can't just do the standard stuff. I think that would be where the CEFC could come in. I think they would have a role to stick around if the government lets them.

Q4

Ben: Connection costs can be a barrier to entry, especially small to mid-sized projects. Assuming increased penetration of RE and increasing grid costs required to meet performance standards (synchronous condensers and reactive plant), how might RE projects overcome this?

To be fair, I haven't gone into it that much, but the stuff I have read is that AEMO are trying to plan for how we get more renewables in the grid along with the TNSPs and so things like the south Australian grid stability you know like this is the kind of stuff they're coming forward with and the networks have to plan five years in advance and part of that is identifying where are the weak spots and what they're going to do to upgrade the network themselves or what needs to be done to upgrade the network and I think that now they're also trying to look at that from a perspective of yeah how do we incorporate more renewables and there was this concept a while ago of kind of the network service providers looking at like hub areas of sort of where in the network there should be renewables and to try and focus in those areas to get the grid connection, you know of how the grid connections could work and to try and encourage people to put their projects there.

Q5

Ben: Renewable energy zones (REZ) have been highlighted in the ISP as a potential solution to reduce costs associated with network infrastructure. Is this likely to be conducive to RE development?

Yeah I don't know that much about it... I guess the only things I can see are it's more just on the ground maybe from a community perspective, they have everybody turning up and building their sites there and I don't know if that's part of that ISP project, which is crap, I haven't read the ISP report yet, but it would be quite interesting to know what they consider… I assume they've looked at it from a resource and grid perspective, but the community aspect with all the other layers and community is one where that whole social licence operate. If the community doesn't want it then you can't really force it upon them. But if they do it right, you can usually manage most of the colder issues and again, it's a mix of when… I'm assuming the space within the area… it's just something to think about and you know I think it would help, although it is just a bit of a free for all, like that's the problem with our system, because people have sites already and if they don't fit within the zones then.

Ben: Ya, that's the thing - it's a bit unfair and potentially good resources as well that might be outside the zones?

Yeah, but maybe if they can raise the money to pay the connection then they can pay for it, then that might be fair enough.

Q6

Ben: The ISP refers to a future energy mix dominated by RE, storage and gas powered generation. How likely is this scenario given high gas prices?

Yeah it's interesting, I was talking to some people in the gas business and they actually think it's not too bad and I guess they're focusing on peaker plants, so fast start.

Ben: So, it would be higher cost and they'd be able to break in because it would be above their operation costs?

Yeah, that's right so they're not needing so much to provide base load… yeah its about covering those high points in the market. From a cost perspective they could probably be paid enough to do that, but difficult – if the cost of gas keeps going up then I'm assuming at some point it becomes uneconomic.

Ben: Is that based on just the price they're getting for their exports? So if its' uneconomical for them, why would they go and sell it locally for less?

Yeah, so I think Western Australia does have a portion of the gas that is reserved for the domestic market so… I don't know whether it's at a set price or whether the domestic market bids for whatever it's going to cost, but there's nothing like that in the eastern gas market, so basically you're competing with export contracts. As I said like we have more gas in Australia than we have capped and at the moment Victoria has a moratorium on gas exploration and so does South Australia, so there is the flip side that if we opened op more exploration then we'd have more supply and whether the government wanted to reserve some gas. It's interesting to think in the future, will it be gas or will some other technology come along and beat them to it?

Q7

Ben: How suitable is the NEM to deal with the future of a significant proportion of RE, given the zero-marginal cost of RE and its effect on wholesale pricing? Is market reform necessary?

I haven't really thought about that question in terms of, once we're at quite a big penetration of renewables that it's not particularly dispatchable how the bidding process would then work. That is a really interesting question that requires a lot more thought that I can really give to it right now.

CONSENT FORM

UNIVERSITY OF CAPE TOWN

CONSENT TO PARTICIPATE IN A RESEARCH STUDY

Research Topic: Large-scale grid-connected renewable energy in Australia: Key success factors

Dear potential participant,

You are invited to participate in a research study conducted by Ben Campbell, masters student at the University of Cape Town. The research study is supervised by Saul Nurick, from the Department of Construction Economics and Management of the University of Cape Town. The results of the research study shall be presented to the Department of Construction Economics and Management in fulfilment of the requirements of the degree in Master of Science in Project Management.

Should you have any queries or concerns regarding the research, please feel free to contact me, Benjamin.campbell@wsp.com. The research supervisor, Saul Nurick can be contacted at sd.nurick@uct.ac.za.

Purpose of the research study

The purpose of this research study is to identify what the barriers and correlating key success factors are for the continued integration of higher levels of renewable energy into the Australian National Energy Market (NEM). This is anticipated to be achieved by carrying out an interview addressing certain key topics that have been identified surrounding renewable energy and the NEM.

Procedure

Your participation in this research study is completely voluntary. Should you volunteer to participate in the research study, we would consult with you in order to agree on a time that would be suitable for a semi-structured face-to-face interview. Various questions shall be asked that will be used to supplement data obtained from documentation analysis under a case study research setup.

Potential benefits to participants

Should you request, the research findings will be shared with you.

Confidentiality

Every effort shall be made in order to ensure that the subjects are anonymous and the safeguard of any information provided. Confidentiality of all information shall be maintained. The information obtained from the interview process shall be used for this research study only. The raw data obtain from the interview shall only be revealed to individuals directly associated to the supervision and marking of this research study.

Participation and Withdrawal

You may elect to withdraw from this study at any time. You may also elect to refuse to answer any question that you do not wish to answer.

Rights of research participants

You may elect to withdraw your consent at any time and discontinue participation without any penalty. This study has been reviewed and granted ethics clearance through the University of Cape Town Research Ethics Board.

Signature of Research Participant/Legal Representative

I have read the information above and my questions have been answered to my satisfaction. I hereby agree to participate in this research study and been given a copy of this form.

....
Name of Participant (Please Pr nt)

....
Company of Participant

....
Signature of Participant